DIGITAL
PHYSICS

DIGITAL PHYSICS

Consciousness is Primary

EDIHO LOKANGA

Printed in the United States of America
First printing edition 2022.

Book covers and design by:
Wordzworth Book Designers and Publishers
www.wordzworth.com

ISBN: 978-1-7396247-0-5

Kwantum Publishing
College House, 2nd Floor
17 King Edwards Road,
Ruislip
London

HA4 7AE

DIGITAL PHYSICS SERIES BY THE SAME AUTHOR:

The Universe Computes, First edition

The Universe Computes, Second edition

The Universe Is a Programmed System

*The Meaning of the Holographic Universe
and Its Implications Beyond Theoretical Physics*

*The Physics of Information, Computation,
Self-Organization and Consciousness Q&A*

Decoding the Universe

OTHER BOOKS BY THE AUTHOR:

Beyond Eurocentrism: The African Origins of Mathematics and Writing

Beyond Eurocentrism: Science and Technology in Pre-colonial Africa (2023)

Cosmology: The Role of Planets in Shaping Our Universe (2024)

*About Love: Messages of Love, Happiness, and the Struggle against Injustice
and Discrimination: Selected Poems*

*Kuhusu Upendo: Ujumbe wa Upendo, Furaha na Mapambano dhidi ya
Dhulma na Ubaguzi (In Kiswahili)*

Bonzambe ya Bukongo: Ngengisa Nzela (In Lingala)

Mabondeli: Bonzambe ya Bukongo (In Lingala)

Masapo: Mobundisi Ye Nde Mobebisi (In Lingala)

Masapo: Bizaleli mpe Bopeto ya Ekelamu ya Tata Nzambe (In Lingala)

Masapo: Bolingo, Bosangani mpe Bosalisi (In Lingala)

DEDICATION

This book is dedicated to my elder brother J. M. Onadikondo Omadjela, a Congolese physicist who left us early this year. His passing has created a void in the family and an unbearable pain that I have not yet come to terms with. It is difficult to accept that he is really gone. I am struggling daily to recover from the loss. A great inspiration to me since I was a child, he guided me and cared for me when I was a little boy. Thank you, brother J. M. Onadikondo Omadjela, for the legacy you have left on this planet.

Brother J. M. Onadikondo Omadjela

CONTENTS

ILLUSTRATIONS

FIGURES

TABLES

ABBREVIATIONS

LQG   Loop Quantum Gravity

QFT   Quantum Field Theory

QG   Quantum Gravity

QM   Quantum Mechanics

SM   Standard Model

ST   String Theory

ToE   Theory of Everything

ACKNOWLEDGMENTS

As with every book I write, I am very excited and filled with joy and contentment, and I always remember with fondness those who have supported me in one way or another and those who continue to help me. As every writer is aware, you live in your own individual universe during writing, a universe filled with various entities both real and imaginary. The names of multiple people surrounding you and those who mean a lot to you will keep coming into your head.

First, I would like to pay homage to my ancestors and some of their descendants still living in Penge Onadikondo, Sankuru, in the Democratic Republic of the Congo. I want to pay tribute particularly to my grandfathers, Chief Onadikondo Omadjela, Chief Ediho Kengete Ta Koi, and Chief Otshudi; to my grandmothers, Mama Esambo Wanya Koi and Mama Akota Mange Ololo; to the Democratic Republic of the Congo and the father of the nation, P. E. Lumumba; and to Mfumu S. Kimbangu, Yaya Vita Kimpa, Chief Ngongo Leteta, etc.

Special thanks go to my father, Chief Lokanga Lopongo Onadikondo, and my mothers, Mama Ehadi Walo, Mama Ombawo Mundeke, and Mama Ohandjo Tendake. Furthermore, I would like to thank my family, namely my wife, Charlotte Lokanga, and my children, Engombe Wedi Lokanga, Shungu Umumbu Lokanga, Maranga Pama Lokanga (Mama Tshike), and Onadikondo Omadjela Lokanga (Papa Tshike). They have always been around, giving me support, assistance, and, above all, the joy to live, the kindest and most beautiful people to have around. May peace and joy be with you forever.

Furthermore, I am grateful for the support and assistance I received from my close friends, colleagues, and relatives in the United Kingdom, the United States of America, Tanzania, Zambia, South Africa, Malawi, etc. Many thanks are also due to Ekumanyi Onadikondo Omadjela, Ekumanyi Lokanga Lopongo "Zulu Zani," Prof. Ernest Parfait Fokoue, Prof. Laurent Cleenewerck, Prof. Rene Yamapi, Dr. Tongele N. Tongele, and Mupe Ndombasi (Ne Ndombasi Kongo).

While writing this book, I read and was inspired by the works of scholars whose names are listed below. Over the years, these researchers have done fantastic work in areas including information theory, consciousness, quantum physics, loop quantum gravity (LQG), and quantum field theory (QFT). I believe this work is crucial in the quest to develop a theory of everything (ToE). Some of these writers' writings have had a significant impact on our understanding of computation and consciousness—with implications for the direction we should give to the field of theoretical physics. From the bottom of my heart, thank you very much:

Peter Russel, Alex Vikoulov, Dr. Bernardo Kastrup, Prof. Art Hobson, James B. Glattfelder, Prof. Carlo Rovelli, Prof. Abhay Ashtekar, Dr. Stephen Wolfram, Kristina de Korpo, Dr. Lee Smolin, Dr. Julian Barbour, Donald D. Hoffman, Prof. Philip Goff of Durham University, Steve Taylor of Leeds Beckett University, the late Prof. David Bohm, the late Karl Pribham, Prof. David Chalmers, Prof. Erwin Laszlo, Thomas Nagel, Prof. Giulio Tononi, Dr. Robert Lanza, Gregory Matloff, Prof. Sir Roger Penrose, Prof. Stuart Hameroff, Prof. Bernard Haisch, Prof. Max Tegmark, Ronald Pokatiloff, Prof. Francesco di Biase, the late Albert Einstein, the late Prof. Max Planck, the late Prof. John Wheeler, Dr. Ethan Siegel, Dr. Julian Barbour, Prof. Seth Lloyd, Prof. Tommaso Toffoli, Dr. Edward Fredkin, Prof. Carlos Gershenson, Prof. John Hagelin, and Dr. Pim van Lommel.

Also, I would like to thank two teachers whom I met at the Coleshill School in Warwickshire in the United Kingdom during my short time there as a physics academic mentor. I briefly discussed the nature of electrons and sought their opinions about the link between matter and consciousness. Thank you to Jodh Panesar, head of the science faculty and the physics department, who gave me his thoughts and opened my eyes to some issues I had not been aware of. Thanks also to Paul Larder, from whom I learned a lot and whom I had the privilege of listening to on the nature of atoms and their constituents. They made me think deeply about what I was researching and reflecting.

My most profound respect and thanks go to Stuart Mitchell, a progressive lecturer, a dedicated master, a true humanist, a guide, a commandant, a gentleman full of heart, an inspirational teacher, and course leader of the Postgraduate Certificate in Education Post-Compulsory Education and Training (PGCE PCET) at Birmingham City University (BCU). He is an authentic human being—the kind of person humanity needs. He understands learners' minds and can guide and inspire you for the rest of your life, a man I will remember for the remainder of my existence, a great soul. He has shown throughout the year the kind of commitment we expect from someone in his position. He is by far the best teacher I have ever met. I have learned so much from him in such a short period—once again, many thanks.

Also, my thanks go to all those who made it possible for me to complete my PGCE PCET with a specialism in science and technology (physics) at BCU: in no particular order, Karen McGrath (program director), Associate Professor Liam McGrath, Dr. Georgina Garbett, Kerry Taylor, and Lynn Kearsey. In addition, I would like to single out Sandra Parsons, who worked extremely hard to get us into placement; thank you very much. Also, Matt Loftus (British mental health guru), an inspirational lecturer and a great educator. His teaching on mental health issues is equal to none. I hope he releases a book (The ABC of Mental Health) to help bring justice in mental health. Furthermore, I extend my gratitude to Amy Dodd, a passionate young lecturer full of love, devotion, and energy, always smiling and engaging, and to Simon F. Whitehouse, the embodiment of peace and love, a wonderful teacher. He makes you fall in love with teaching. He instructed us about Ofsted and other important educational issues. And finally, to Dariusz Uzarewicz (administration department), an important player in the school of education, many thanks for the opportunity you gave me.

I am forever grateful for your kindness, support, and understanding and for the love we gave one another, my friends and colleagues of the 2021–2022

PGCE PCET at BCU. A wonderful group of human beings, full of devotion and kindness, you have made this year a very special one to remember. With adoration, affection, and admiration forever, I send my love to all of you. You will always have a special place in my heart. Many thanks (in alphabetical order):

Aisha Din, Aishah Malik, Amelia Whittle, Andrew Bending, Andy Tatman, Caleb Madzokere, Cas Rooney, Casey Hayden, Colin Vernon, Daisy Wood, Dean Avis, Deborah Walker, Ellie Mousley, Elson Sibanda, Emily Kite, Emma Knight, Enakshi Dasgupta, Hannah Moyle, Hannah Palethorpe, Harriet Berrington-Hughes, Hennah Ashraf, Holly Wagstaff, Ismail Ahmed, James Barton, James Bradshaw, Jordan Walker, Dr. Kevin Buckland, Laiba Madjidzadah, Lauren Flanagan, Madeleine Scott, Malik Umar, Matt Ward, Matthew Spurrier, Mohammad Khan, Mohammed Ali, Mohammed Khan, Mohammed Miah, Morgan Whitebeam, Muhammad Ali, Dr. Muhammad Khalid, Naina Begum, Natalie Sparrow, Osama Nasir, Rebecca Williams, Rhiannon Bunce, Rushna Begum, Saarah Fatima, Sam Forge, Samantha Harley, Sameeya Khan, Samuel Ross, Sara Khan, Shamila Ikhlaq, Shannon Luscombe, Sharna Emmett, Sifiso Magongo, Siobhan O'Shea, Tamara Martin, Wadia Yafai, and Zahra Baig.

Besides, I would also like to pay a great tribute to the new generation of Pan-Africanists carrying the torch and leading the struggle to free Africa and the world from the bondage of imperialism, capitalism, and exploitation. It is incredibly wrong that humanity is controlled by a small group of people who have accumulated most of this planet's wealth. There is so much suffering, injustice, and poverty on this planet. It is time that we all stand against oppression, capitalism, and all forms of exploitation. Once again, I salute the emergence of a decisive new Pan-Africanist leadership opposed to any form of exploitation. The new direction is inspiring a new generation of youth worldwide, with a vision to bring a new way of governance, free from exploitation and imperialism, with the sole aim of serving all its people in political, economic, and racial equality.

My respect to you Justin B. Tagouh (director general of Afrique Media), President Banda Kani, Patient Parfait Ndom, Dr. Bertrand Tatsinda, Dr. Yamb Timba, Christel Andre Fanga, Prof. P. L. O. Lumumba, Madame Manuela Sike,

Paul Ella, Francois Bikoro, Madame Amina Fofana, Guy Nfondop, Stive Jocelyn Ngos, Kemi Seba, Professor Nioussérê Kalala Omotunde, Hubert Etoudi, Dr. Patrick Sappack, Henri Diabate Manden, the astrophysicist Prof. Jean Paul Mbelek, etc. We are very grateful for the work you are doing. Thank you very much. We are all behind you. Keep up the excellent work.

PREFACE

Most physicists take it for granted that the universe is made up of matter; that matter, in turn, is composed of atoms; that atoms made up of particles, such as electrons, protons, and neutrons; that protons and neutrons themselves are made up of quarks; and that everything in nature is governed by the known laws of physics and chemistry. I only partially share this point of view. I firmly believe that many phenomena in the universe depend on rules or factors not yet incorporated into the physical sciences. The last few years have led me to reflect on the many unsolved problems in physics, such as the quest for the theory of everything (ToE), the arrow of time, the interpretation of quantum mechanics, and the fine-tuned universe, to mention just a few. This state of affairs made me think and pushed me to do something about it.

I realized that something was missing in our quest for the ToE and in our understanding of physics. I then became more determined to write this book in order to explain how the conventional approach is too restrictive. Since I was a little boy living in the Democratic Republic of the Congo, I have believed that consciousness is primary. I am partly influenced by my background: Africans generally posit that consciousness is a creative force that shapes the universe or nature, including matter, and is the source of everything. I saw an opportunity, an open window, to outline a new way of looking at physics. Likewise, with this firm conviction in mind, I saw an opportunity to outline a new way of looking at physics by incorporating consciousness into the physics equation. Like many new hypotheses, it is essentially speculative, and like any other, it will have to be tested experimentally.

The purpose of this book is first and foremost to explain the fundamental role of consciousness in physical science and its role in running the universe, including in relation to matter and particles. I am firmly convinced that it is simply wrong and implausible to deny consciousness its causal role. The following pages aim to establish its prominent position and its meaning, insisting on the idea that consciousness is fundamental. It follows that the lack of progress that has been made in the quest for the ultimate theory of nature is partly due to the limitations of trying to understand everything in terms of matter and energy. However, suppose everything—including matter, energy, life, and mental processes—is described in terms of information and consciousness. In that case, much progress can be made in the search for the ultimate theory of the universe.

For several years now, theoretical physicists have been working on the ToE or grand unified theory. The most "successful" candidate, string theory or superstring theory (ST), attempts to unite all the forces of nature: electromagnetism, strong and weak nuclear force, and gravity. There are many shortcomings to this model, despite the progress that has been made over the years. For instance, it has not so far stood up to scientific experimental scrutiny. As a result, it cannot be seriously considered as a contender for a ToE. A competitor in the form of LQG is also making progress and has achieved notable successes. However, to realize the dream of a ToE and the theory of quantum gravity (QG), we must develop a new approach that combines the virtues of these two theories with an emphasis on consciousness.

Let me stress once again, as brilliant and successful as ST and LQG have been over the last twenty years, they must not be viewed solely in terms of matter and energy. One additional component— called consciousness—is needed to unlock their secrets. If we are to make progress, we must incorporate consciousness as an ingredient of physics. These two theories were built without mentioning the role of consciousness. I firmly believe that consciousness is an essential and fundamental missing ingredient in both theories.

We learn from QFT that subatomic particles are not physical objects or discrete entities but rather fields. Particles are continuously in a state of vibration. An elementary particle is a vibration of an underlying quantum field. The particle is the field. This underlying field is intelligent. What the proponents of QFT call a field, I refer to as an intelligent system or a smart field. Why? The field carries information, performs computation, self-organizes, and is creative. In a nutshell, one of its most important properties is the ability to create. It is imbued with a creative force. The whole process hints at the presence of intelligence or consciousness. Matter, in its original nature, is consciousness. It is a form of consciousness that materializes and can be perceived through our senses.

INTRODUCTION

For several centuries now, science and technology have made unbelievable progress in lifting humankind's social and economic status on this planet from poverty and obscurity to a better situation. Its many achievements encompass anatomy, biology, physics, chemistry, astronomy, and psychology. These achievements are not worth repeating, as they have been covered in several of my books. Despite the progress that has been made, one outstanding issue is the understanding of consciousness. Therefore, I would like to discuss consciousness—its role in physics and in understanding the universe—in the sixth book in the *Digital Physics* series. This manual is not about a new theory of knowledge; it is merely a treatise discussing the critical role that consciousness must occupy within physics if the latter is to make concrete progress in its quest for the ToE.

Various books and articles presenting different theories of consciousness have been written by eminent researchers, such as David Chalmers, Erwin Laszlo, Peter Russel, Daniel C. Dennett, and others. I must admit that consciousness is an issue that I have avoided in the past whenever it has crossed my mind. As a physicist, discussing the non-material is not part of my academic upbringing. However, during my research into the field of digital physics (for my work *The Physics of Information, Computation, Self-Organization, and Consciousness*), it kept arising in various forms. I felt that something, somewhere, was missing in physics that would unlock the ToE. That is why I think it is essential for scientists to pay attention to consciousness. I am saying that incorporating consciousness in science will resolve several important issues regarding our existence on this planet.

Over the last fifty years or so, consciousness has been a recurring topic in physics and science in general. Along with matter, energy, and information, consciousness has started to occupy an essential role in physics. Although not all physicists agree about its role in physics, nonetheless, it is there. A minority within the physics community thinks that the failure to include consciousness in scientific models of the universe has impeded the search for the ToE. However, the majority such as those working in the field of ST or LQG take the opposite view. ST is a theory that replaces subatomic particles with strings. The strings are either closed or open loops. They vibrate in different ways, and the different modes of vibration give rise to all the different particles in the universe. One the other hand, LQG is one of the candidates for a quantum theory of gravity. Loop quantum gravity is based on classical general relativity. One of the essential conclusions of LQG is that gravity should be quantized. Both theories do not see the need to involve consciousness in their work. This is a mistake because the role of consciousness was without a doubt stressed by the pioneers of QM, who ascribed a central role to it.

Several centuries ago, the existence of consciousness was not questioned. It was accepted in the same way as the existence of our physical world. However, in the years following the Industrial Revolution, a materialistic view of the universe took precedence. And over the years, most QM interpretations removed the consciousness element from their frameworks. Various hypotheses were proposed that did not include consciousness in their explanations of the nature of our universe: the theory of evolution, first proposed by Darwin, and the Big Bang theory, so dear to physicists, are two examples. Since then, despite the progress that has been made, several issues and questions have not been answered. This reductionist approach is being answered by a new wave of scientists arguing for consciousness to be given a fundamental role.

The materialistic view that puts matter at the center and assumes that our brains produce consciousness has lost ground. This is because several years of theoretical and experimental research have so far not shed any light on this issue. This is despite efforts to understand how our brains work. Those working in neuroscience, the scientific study of the nervous system, have found some mismatches between consciousness and brain activity. This view is not different from that of the overwhelming majority of scientists and philosophers, who adhere to the school of thought, according to which the universe started with a Big Bang, meaning that the universe was created mechanically and can be explained mathematically, without the need for consciousness. As shown in my last book, *Digital Physics: Decoding the Universe*, this model is plagued by many problems. This development has led some leading researchers in the field of consciousness, such as Thomas Nagel, David Chalmers, Giulio Tononi, and Robert Lanza, to cite just a few, to reject the idea that the brain directly produces consciousness.

The view that the brain produces consciousness is difficult to account for. This approach cannot explain the full range and depth of consciousness. For instance, the location of consciousness within our brain has not been pinpointed. This realization is pushing a new wave of researchers to look for alternative explanations. On the one hand, some scholars believe that the brain does not produce consciousness. On the other hand, some suggest that consciousness is a field around us, like the wind or the sun, that is transmitted into our brains. The third school of thought posits that consciousness is embedded inside everything in the universe, even though its location is not yet precise.

If consciousness is taken as a fundamental entity, it becomes the source of everything. Those who believe that consciousness is everywhere are talking about the concept of "fundamental consciousness," or "proto-consciousness," a field that extends through all of space. One proponent of this idea is Gregory Matloff (2017). I firmly believe that the idea of consciousness as the source of everything, a fundamental quality of the universe, can answer several questions. For instance, why do we experience oneness? We experience a sense of unity because we share this fundamental consciousness. Our minds are connected. The mind is a subtler and fuller expression of consciousness. The idea of consciousness as an essential quality of the universe offers us a platform to understand the world in a new way.

Many scientists are arriving at similar conclusions. For instance, the work of the British mathematician Sir Roger Penrose, in collaboration with Professor Hameroff, which uses quantum theory to explain the phenomenon of consciousness, is a revelation. Their joint research may have discovered the carrier of consciousness (taken here as information or elements that accumulate during one's life). At death, consciousness moves somewhere else. This position has also been taken by the German physicist Bernard Haisch, who has suggested that quantum fields permeate all empty space (quantum vacuum) and are responsible for producing and transmitting consciousness (Powell 2017). This attempt to explain the universe in a new way changes our view of our physical world and our place in it and asks us to confront pressing questions and issues facing humanity.

A host of questions that we must answer about the new view arise. Although it is clear that some of these questions are difficult to answer, this book will look at them attentively. And, inspired by the pioneers of quantum theory, the theory of wholeness, and the concept of quantum vacuum or zero-point energy (ZPE), we can use their work as a starting point to learn about consciousness and thus make some progress. To start with, let me first remind you of the views of the founders of quantum theory. The pioneers of quantum theory gave a central position to consciousness. Consider, for instance, the following statement:

> *Max Planck: "I regard consciousness as fundamental. I regard matter as derivative from consciousness" (quoted in Sullivan 1931).*

> *In addition, as rightly pointed out by Sunderland (2015):*

> *Von Neumann: "Consciousness, whatever it is, appears to be the only thing in physics that can ultimately cause this collapse or observation."*

> *Freeman Dyson: "At the level of single atoms and electrons, the mind of an observer is involved in the description of events. Our consciousness forces the molecular complexes to make choices between one quantum state and another."*

Eugene Wigner: "It is not possible to formulate the laws of quantum mechanics in a consistent way without reference to the consciousness."

Pascual Jordan: "Observations not only disturb what is to be measured, they produce it."

Niels Bohr: "Everything we call real is made of things that cannot be regarded as real. A physicist is just an atom's way of looking at itself."

Wolfgang Pauli: "We do not assume any longer the detached observer, but one who by his indeterminable effects creates a new situation, a new state of the observed system."

Furthermore, Max Tegmark: "I believe that consciousness is, essentially, the way information feels when being processed" (2007)

These statements, coupled with accumulated scientific evidence over the last few years, are forcing scientists and philosophers to reconsider the explanations of consciousness as a product of the brain. However, the idea that consciousness is a separate entity distinct from physical matter is gaining ground. The faculties of mind, intellect, and personality are simply manifestations of consciousness. The nonmaterial part of us is consciousness, soul, or spirit, and the body is simply the means through which consciousness expresses itself and experiences the world.

Spirit, consciousness, inner self, anima/animus, life energy, essence, or "I" are mostly synonyms for the word "soul." Consciousness operates throughout the body. It appears that it is not the physical body that is responsible for our actions but the consciousness or thinking being that forms the body's operating system. Very simply, it is I, the self, or the soul. The soul uses the word "I" for itself and the word "my" when referring to the body: my hand, my mouth, my brain, and so on. I am not the same as my body. The idea that the entire universe is conscious or that consciousness permeates everything is gaining ground all over the world.

This new understanding of consciousness suggests that the physical universe and the laws and properties of matter, energy, etc. are the physical manifestations of consciousness. As you can see, I have brought the concepts of matter and energy into the equation. Why? For some researchers, we must revise our views of what matter and energy are because the old way of seeing matter as a physical object made up of particles is a stumbling block for further progress. A new generation of researchers is looking closely at the role and meaning of information and questioning our understanding of matter and energy. For instance, those who favor the concept of information argue that underneath the levels of atoms, protons, electrons, quarks, etc., matter or the universe is made up of binary digits (bits) or quantum bits (qubits). A bit has a single binary value: zero or one. A qubit is a quantum bit. In quantum computing, this is the basic unit of quantum information, and it is the quantum version of the classical binary bit that is physically realized with a two-state device.

Others also use information to explain the universe but give a more significant role to another, new, element, namely consciousness. They see consciousness as the source of everything, as the force behind energy, matter, and the organization of the universe. For them, consciousness is fundamental. There is a need for a new direction in physics. Our leading theories, such as ST, have not been able to achieve their expected aims. That is why a search for a fresh perspective is embodied in this book. This text ascribes a more significant role to consciousness. It sees consciousness as an intelligent entity that holds everything in the universe together and is responsible for running our physical bodies. Consciousness can be seen as the creative force, the life force embedded in everything in the universe.

The double-slit experiment is one of the most famous physics experiments. The results of the investigation show that light and matter can display the characteristics of both classically defined waves and particles. They also reveal the fundamentally probabilistic nature of quantum mechanical phenomena. The outcome of this experiment suggests that there is this invisible, omniscient energy—intelligence—embedded in electrons. The experiment has led to much discussion about what matter, and energy are. More and more people are digging deeper to join the search for the missing pieces of the puzzle because they sense that something is missing in the quest for the ultimate ToE. The mystery of matter has not been solved, and this raises eyebrows and challenges some of the scientific community's best minds.

Experiments in QM have demonstrated beyond doubt that consciousness plays a fundamental role in every operation. For instance, factors associated with consciousness, such as observation, measurement, thinking, and intention, directly correlate with our physical world, as pointed out by several physicists. The quantum double-slit experiment demonstrates how our consciousness and the physical world are intertwined, making us realize that the observer creates reality. This experiment (see Figure 0.1) has played a crucial role in helping us to understand how consciousness shapes the nature of our physical world or reality. Although the scientific community has not yet come up with credible explanations for the extent of this connection, the conclusion is clear that consciousness is fundamental, *the source of everything*, and that electrons are imbued with consciousness.

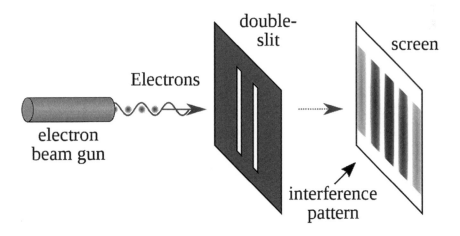

Figure 0.1 *An illustration of the double-slit experiment. Photons or particles of matter produce a wave pattern when two slits are used. Image by NekoJa NekoJa via https:// commons.wikimedia.org/wiki/File:Double-slit.svg (Creative Commons CC-BY-SA-4.0 license)*

This experiment shows the relationship between matter, energy, particles, waves, and consciousness. Many physicists have demonstrated their unwillingness to deal with the concept of consciousness in their experiments. They may have been influenced by the reluctance of Isaac Newton to define the primary source of gravity. "Hypothesis non fingo," he replied, which translates as, "I do not have a clue." What is concealed in matter is the expression of consciousness, which is primary. Therefore, it is important to stress that Newton's physics was constructed without solving two fundamental problems: the causes of attraction and the workings of consciousness. Physics and chemistry tell us little or nothing about the intrinsic nature of matter; they only describe for us what matter does. However, we suspect there must be more to the nature of matter and physical entities.

Physics describes everything in terms of matter, energy, space, time, and information. For the majority of physicists, the universe is made up of matter. Physics and chemistry do not tell us about the intrinsic nature of matter—what matter *is*. We only learn from physical sciences about the behavior of matter (i.e., what it does) and some of its properties, such as its mass and charge. It behaves in a certain way; for instance, it obeys the laws of gravity, resistance, attraction, and repulsion.

The constituents of matter—atoms, molecules, electrons, quarks, elements, etc.—are intelligent entities. For instance, when two elements combine, they must have the intelligence to create something. Atoms combine to form molecules. An atom is made up of a vast number of elementary particles. These composites can only come together by the creative intelligence embedded in them. The concept of consciousness as primary suggests that atoms are intelligent entities imbued with consciousness. For instance, an element such as sodium is composed of many elementary particles all imbued with consciousness. It will produce a more complex system, a molecule, if it combines with an atom of another element, such as chloride. All known elements hold information; they are creative and interactive.

Experiments in QM, particularly those examining the behavior of electrons, give us a clue as to what matter is made of. It looks likely that matter is imbued or embodied with consciousness. Consciousness is the intrinsic nature of matter. Thus, by looking at the behavior of matter (i.e., what it does), we learn about its action in the physical world, and we also understand its intrinsic nature. By looking at it closely, through QM experiments, we remember and start to appreciate its inherent nature. Thus, we get a better view of matter and the universe.

Although the main difficulty for some might come from the fact that consciousness is invisible to our naked eyes, I would like to stress that QM experiments show that the universe's fundamental constituents, such as electrons, have simple forms of experience. This approach gives a platform in which to integrate consciousness into physics and science in general, thus giving matter a new status. The description of matter as made up of atoms, electrons, particles, and elements brings to mind a physical, tangible object when, in fact, it is not. This realization leads us to ask a host of questions. If particles are not material objects, what are they? And if there are no particles—if these are just models that physicists use—what is really out there?

Consciousness is the basis of reality. Every human being experiences consciousness on a personal level. Consciousness is required so that the universe materializes in front of us or in our own minds. Consciousness is fundamental. Without it, there is no life or matter, and living entities cannot move. Without consciousness, the body is dead; it will just collapse. Consciousness is a force within the universe and within the physical body. When consciousness leaves the body, the latter is finished. Everything starts with a perception; we see things through our minds (consciousness). Day-to-day existence clearly shows that our thoughts, inventions, images, feelings, etc. are where everything starts.

When someone or somebody has the experience of seeing something, such as a tree, what they see is not the tree itself (matter), the physical object; rather, it is the image that appears in the mind of the person. This is to say that everything around us, everything we perceive—feelings, colors, sounds, sensations, etc.—is first and foremost a form appearing in the mind. Thus, it is vital to distinguish the physical objects we see and the images in our minds. The pictures in our minds are not to be confused with the material or external world; that would be a mistake, an illusion. An object, such as a table or my computer, is made up of matter, which is made up of atoms, molecules, electrons, quarks, bits, or qubits.

On the other hand, the image in my mind is not made up of the same stuff as my table. Peter Russel clearly explains this:

> The English language does not have a good word for this mental essence. In Sanskrit, the word chitta, often translated as consciousness, carries the meaning of mental substance, and is sometimes translated as "mindstuff." It is that which takes on the mental forms of images, sounds, sensations, thoughts, and feelings. They are made of "mindstuff" rather than "matterstuff" (2006).

He insists on the ability of consciousness to take on the form of every possible experience.

In *Digital Physics: Decoding the Universe*, I discussed dreams and showed how the images we see are composed of the same stuff or entities as our dreams. The mental substance from which our experiences are formed comes from "mindstuff" or the ability of consciousness to take on form. Along the same reasoning, everything we see, animate or inanimate, emits a specific amount of energy. The emitted light or energy comes with a particular frequency. If one changes the frequency, various colors can be seen in the mind. An object by itself has no color: it is the frequency of light that exists as an experience in the mind.

Additionally, space and time in physics are just a dimensional framework in which the mind constructs its experience. Einstein showed that space and time have no absolute objective states. It appears that space, time, matter, and energy have no fundamental objective status. Physics must use the model to explain the universe, but that model must not be confused with reality. The physical world is not like the forms that appear in our consciousness. Similarly, the forms that appear in our consciousness are not like the physical world out there.

It is clear now that the models that we use in physics, chemistry, biology, etc. to explain the universe do not consider consciousness or how our daily lives are constructed in our minds as a form appearing in consciousness. Mental images are not made up of physical solids (objects, substances) but rather of "mindstuff." This emerging worldview, heralded by Erwin Laszlo, Peter Russel, etc., considers consciousness as the source of everything that exists. It is backed by advances in QM and is not contradicted when one looks very closely.

In this book, an attempt has been made to highlight developments in the last few years supporting the view that consciousness is primary, the source of everything. The idea is destined to have an important place in scientific research. There is no doubt that this emerging view is causing a paradigm shift in our understanding of our physical world and the universe. Furthermore, this guide stresses the important role that consciousness plays in the universe. It explains the mysteries of matter, energy, space, time, and perception. This new approach to the study of consciousness can solve numerous problems in the field of theoretical physics, which do not yet have any solution.

Those readers who are looking for an alternative to the materialistic paradigm and want to learn about consciousness as the driving force of the universe will find this book extremely useful. If one wants to study consciousness, one must realize that the study of consciousness is another science. One should look at it from another angle. It is not a science of physical objects, like chemistry or physics nor is it the science of electrons and photons. It is a science that opens the door to incredible knowledge and answers some of our fundamental questions. For instance, who are we, why are we on this planet, is consciousness a separate entity from matter, etc.?

Finally, I would like to stress once again that this manual is not a book about theories of consciousness. Many books about consciousness have been written. Here, the emphasis is on the fundamental role that consciousness plays in the universe. By the end of this book, the readers will have learned why consciousness is so fundamental; they will appreciate the book's main idea and its approach to understanding consciousness, matter, energy, space, time, and perception in a new framework.

In addition, the concept of energy is clarified. I explain why I consider the study and understanding of consciousness to represent the best path to resolving many puzzles and to arrive at a ToE. I stress the fundamental role of consciousness and argue that it is the source of everything, the divine force running the universe. After this introduction, I discuss various concepts to emphasize the meaning of consciousness and its role in physics and the universe.

Therefore, this book is divided into six chapters, plus a glossary that defines scientific terms used throughout the book.

In Chapter 1: The Mystery of Matter, I argue that despite the progress that has been made in the study of matter in various scientific fields, there is still much that we do not know about this concept. That is why I begin with a discussion of the composition of matter and, above all, its intrinsic nature, citing research from various fields of physics, including QM, QFT, ST, information, computation, self-organization, and consciousness, to reveal a new picture that many in the physics community tend to ignore. A new conception of matter is proposed, and explanations are put forward.

In Chapter 2: The Illusion of Energy, the intrinsic nature of energy is probed. Here, I discuss the mystery and origin of energy. What is energy? Where does it come from? Is energy an abstract concept? Has anybody seen any form of energy? I discuss the Casimir effect, the ZPE, and the possible connection between energy and consciousness and its implications. I stress that energy is a potential matter that has yet to solidify. I hypothesize that emotions are forms of intangible energy released by bodies.

Chapter 3: Revising Our Concepts of Time, Space, and Perception is essential in making us think once more about the meaning of time, space, and perception. These are concepts we take for granted, but when we look closely at these concepts through the angle of consciousness, astonishing revelations are laid out. I start with a discussion of the everyday meaning of time and move on to look at the view of two physicists: Prof. Carlo Rovelli's view of time and Dr. Julian Barbour's view and philosophy of time. Furthermore, I look at the meaning of time and perception in various branches of physics.

In Chapter 4: The Role of Information in Understanding the Universe, different types of information are probed and analyzed in relation to matter and energy. Moreover, I discuss the laws of information and their meanings. I answer several fundamental questions, including how information can contribute to our understanding of consciousness. Furthermore, I elaborate on the creative process of information and the role of consciousness in natural systems.

In Chapter 5: The Universal Field of Consciousness, I build on the previous chapters and the evidence presented throughout the book and explain why I think that consciousness is fundamental. In a nutshell, a review of various papers leads to a postulation that consciousness is a quality or attribute of quantum-like processes. In addition, new theories of consciousness posit the existence of the so-called unified field of consciousness. I discuss, among other things, the electromagnetic theories of consciousness and quantum unified field theories of consciousness.

Chapter 6: Conclusion emphasizes the role of consciousness in understanding physics and the universe. I summarize why consciousness is fundamental, stressing that consciousness is primary, the driving force of creation.

I have added a glossary and defined various technical words to help readers to understand any new terms or concepts they encounter in the book. I have avoided excessive technicality and unnecessary details, as the emphasis is on establishing a unique perspective, i.e., the fundamental role of consciousness in the universe and physics. Dear readers, enjoy your reading.

Ediho Kengete Ta Koi Lokanga,
Tipton,
West Midlands,
The United Kingdom,
19/05/2022

PART ONE

———

The Mystery Of Matter

What is Matter?

Concerning matter, we have been all wrong. What we have called matter is energy, whose vibration has been so lowered as to be perceptible to the senses. There is no matter.

−ALBERT EINSTEIN

We learn from physics and chemistry that everything in the universe is made of matter and that matter, in turn, is made up of atoms. An atom is composed of a nucleus that contains protons and neutrons and is surrounded by electrons on the outside. There are four well-known states of matter: solid, liquid, gas, and plasma. This means that all matter, whether solid, liquid, gas, or plasma, is composed of atoms. The only difference between the four states of matter is the arrangement of electrons inside them. For instance, gas is made of the same fundamental particles as solid matter; it is merely the arrangement of these particles and the force that binds them together that makes it a gas. Is this true? Let us together find out the veracity of this statement.

To find out what matter is, I would like to discuss it in the context of QM, ST, QFT, and the new digital physics field. First of all, let me start with what QM tells us about matter.

Quantum Mechanics

We learn from QM that the mass of an atom is concentrated in the nucleus. But there is so much space inside an atom between the nucleus and the atom's edge. An atom is 99.99% empty space. There is also an open space between the nucleus and the electrons surrounding it, so it is easy to compress matter. Although physicists speak of "emptiness," this so-called space inside atoms is not in fact open; rather, it is filled with a fluid of invisible virtual particles. An unlimited space within matter exists. Within these spaces, there are infinitely small particles revolving at high speed and bound together by forces.

When we look at an atom through a powerful microscope, we find that there is nothing there that we can call the "building block" of matter. These tiny particles have no dimension; they are dimensionless; they are not physical objects. It is precisely here that the picture becomes a bit complex. Something is holding everything together; this something is energy; without it, things would not stay together. This solidity that we see with our naked eyes and call "matter" is thus an *illusion*. There is no solidity at all; particles are not solid objects; there is only energy or pattern of vibrations.

The term "illusion" is the wrong one to use, however. Why? The use of this word has led to a recent trend among some scientists and the general public to suggest that matter or the world in which we live is an illusion. One of the issues with this attitude or view is that it leads to an indifferent attitude towards matter, society, the economy, and life in general. Let me make it clear that the idea that everything physical is merely an illusion is misleading, though, at the same time, it is appealing to many people, as it gives them an excuse to avoid facing their problems. This attitude has led to hostile, negative attitudes, neglect of duty, etc. In the following pages, I explain why the word "illusion" is wrong and misleading. For instance, when physicists use this word, they simply mean that we only witness a part of reality and have an incomplete view of our world or the universe.

As rightly pointed out by Steve Taylor (2017), this mistake stems from the translation of the Hindu concept of Maya. He explains, "Maya is the force that deceives us into thinking of ourselves as separate entities and the world as consisting of separate, autonomous phenomena. In other words, maya prevents us from seeing the world as it really is. It blinds us to the unity that lies behind apparent diversity. It stops us from seeing the world as brahman, or spirit. So, it doesn't literally mean that the world is an illusion, but that it's not as it seems."

In the following lines, I would like to present evidence from QM that clearly shows the connection between matter and consciousness and its profound implications for humanity. This knowledge is crucial in shifting the world-view and transforming society through understanding that consciousness is primary, that it is the source of everything, and that matter is not a phys-ical object. Hopefully, this realization will have the potential to transform humanity and lead to a new, peaceful world where we understand every living being's place and take care of each other.

Atoms (see Figure 1.1) and their constituents (particles electrons, quarks, etc.) obey QM laws. These things are so small that they do not adhere to classical mechanics (the physics that describes the motion of macroscopic objects). To explain the concept of matter, let me briefly discuss some of the most prominent features of quantum physics. It is important to stress that things in the quantum world do not behave in the same way as solid objects with respect to their interactions. The picture that we have of an atom sur-rounded by a nucleus, with protons, electrons, quarks, etc. zooming around a neutron is an incomplete one. It is just a model that helps us visualize and thus understand what goes on inside matter.

However, looking at matter's structure very closely through a powerful microscope, you will not see any physical particle (i.e., any physical object) but rather relatively small, invisible vortices (small energies). The deeper you focus, the more you see that there is nothing, just a physical void. What does it mean? It merely means that an atom is not a physical object; it has no physical structure, leading to the realization that atoms are made of invisible energy, not physical matter. It is now clear that QM demonstrates that atoms are made of vortices of energy that are constantly popping into and out of existence. If we look closely through a powerful microscope and analyze various QM experiments and ongoing experimental works with the Large Hadron Collider, our view is slowly changed.

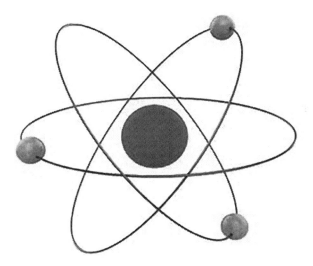

Figure 1.1   *An atom with a nucleus, surrounded by three electrons*

The subatomic particles that make up the atoms or matter are not physical objects. They have no structure or size nor do they have a physical presence. Because they are not physical objects, they have no height, no mass, and hardly any width. In a nutshell, they have zero dimensions. They are just events in time. Table 1.1 shows the electric charge, atomic charge, mass, and atomic mass of some subatomic particles. Quantum Mechanics tells us that to understand reality or our physical world, the universe, we should no longer think of the universe as a physical object in which what we see, sense, and touch are all that exists.

Table 1.1   *Electric charge, atomic charge, mass, and atomic mass of some subatomic particles*

Particle	Electric Charge (C)	Atomic Charge	Mass (g)	Atomic Mass (AU)
Protons		+1	1.6726	1.0073
Neutrons	0	0	1.6740	1.0078
Electrons		−1	9.1094	0.00054858

The point is that when one looks deeply at an atom through a microscope, there are no physical objects. Still, various operations are taking place—computation, movement, etc. All these so-called particles are not made of any solid substance. Observation under the microscope suggests that there is hardly any substance. Therefore, a shift in our understanding is needed. The reality is that when the general public believes that physicists have figured out everything about matter, experiments involving matter are coming out to teach us new things. It is now clear that we cannot see the components that make up matter with our naked eyes. Matter is made of things we cannot see and are not physical.

The non-physical property of matter is behind the force that is running the universe. That force is consciousness, which plays a significant role in our reality's physical makeup. Even scientists themselves are bewildered about the relationship between matter and consciousness. Some have even gone so far as to suggest that matter is a conscious product because merely observing and analyzing the double-slit experiment affects the behavior of matter. In other experiments, too, consciousness has been shown to be capable of altering the nature of matter and energy. As rightly stressed by Max Planck in his 1918 Nobel Prize acceptance speech:

> As a man who has devoted his whole life to the most clear-headed science, to the study of matter, I can tell you as a result of my research about atoms this much: There is no matter as such. All matter originates and exists only by virtue of a force which brings the particle of an atom to vibration and holds this most minute solar system of

the atom together. We must assume behind this force the existence of
a conscious and intelligent mind. This mind is the matrix of all matter.

And in January 1931, he made a similar statement in an article that appeared in *The Observer* when he was interviewed by J. W. N. Sullivan: "I regard consciousness as fundamental. I regard matter as derivative from consciousness. We cannot get behind consciousness. Everything that we talk about, everything that we regard as existing, postulates consciousness" (Sullivan 1931).

Another fundamental point about the nature of matter is summarized by Erwin Schrodinger, one of the founders of quantum theory:

What we observe as material bodies and forces are nothing but
shapes and variations in the structure of space. Particles are just
schaumkommen (appearances). The world is given to me only once,
not one existing and one perceived. Subject and object are only one.
The barrier between them cannot be said to have broken down as
a result of recent experience in the physical sciences, for this barrier
does not exist. (quoted in Vallabhajosula 2009, 59).

Physicists have to use the words at their disposal to convey information, explain the universe, etc. Sometimes, the true meaning of the concept gets lost in more detailed comments. Quantum Mechanics does not tell us that the world is an illusion. My essential point is that, unfortunately, despite the theoretical and experimental findings of QM, many physicists and scientists today still cling to the materialistic worldview. The question is, why? The results are evident in black and white. Matter is not physical; we human beings are not material; the world is not physical. Quantum theory must not be restricted only to the subatomic world and must extend to everything in the universe.

I have shown that matter in itself is convertible into energy or consciousness. Are we humans only made of matter? The answer is clear: no. Understanding matter in light of QM answers several questions and opens up a new avenue for understanding who we are and why we are here on this planet. We must abandon the notion of a materialistic universe, bring an end to the reductionist approach (the tendency to reduce physical phenomena to their most essential parts), and embrace the concept of wholeness (everything is connected and continually interacting). Quantum Mechanics teaches us that there is a fiery spark of consciousness latent in everything in the universe, however insignificant.

Physics, biology, and chemistry in their present forms tend to overlook the role of consciousness in nature and the universe. Consciousness is an integral part of matter or nature. It is inextricably interwoven with all its many transformations (such as growth or changing state of matter). Energy—i.e., consciousness—is matter in its subtlest sense. There is a saying among the Bantu that everything that is created started as a form in the mental universe before it manifested in the physical or material world. Scientists and particularly particle physicists are today realizing the complex intelligence embedded in every single atom, particle, and cell in the universe.

Those energies, whose forms appear as a physical structure in our physical world, can now be understood in a new light. Hopefully, these discoveries reveal the fundamental mistakes humanity has made in thinking that matter is made of solid stuff. Hopefully, it will lead to a radical change in our view of the world, particularly in the West. Quantum Mechanics forces us to accept that the basis of our physical world is non-physical, non-material, and does not consist of particles or solid objects but non-material forms. Commenting on these forms, Diogo Valadas Ponte and Lothar Schäfer said:

These forms are real, even though they are invisible, because they have the potential to appear in the empirical world and to act on us. They form a realm of potentiality in the physical reality, and all empirical things are emanations out of this realm. There are indications that the forms in the cosmic potentiality are patterns of information, thought-like, and that they are hanging together like the thoughts in our mind. Accordingly, the world now appears to us as an undivided wholeness, in which all things and people are interconnected, and consciousness is a cosmic property (2013, 602).

Strange as it seems, QM's conclusions and implications are far-reaching and need to be considered in all spheres of life. After learning the basics of matter, atoms, and subatomic particles through the QM viewpoint, one can safely say the following about its meaning and implications:

- Reality is not determined by physicality.
- The observer creates reality.
- The beliefs, perceptions, thoughts, and attitudes (consciousness) create reality.
- The universe is immaterial.
- Subatomic particles are not physical objects.
- Everything in the universe is connected; that is why information exchange is done instantly.
- Matter is not an illusion, but the atom that makes it up is made up of space, information, and oscillating condensed energy. This simply means that an atom is made of space, forces, oscillation, and information imbued with intelligence.
- Hardness is a property of matter; it is just the power of electromagnetic bonds between molecules.
- We can see less than 1% of the electromagnetic spectrum and hear less than 1% of the acoustic range.
- An atom is 99.99% empty space; therefore, one can compress matter easily, such as gas and liquid.
- The double-slit experiment shows that matter sometimes behaves as particles and sometimes as waves during investigation.
- The observation of a particle now may affect what happened to another in the past. The present affects the past. This is known as Wheeler's delayed-choice experiment or quantum eraser.
- A particle can be in two places at the same time. This is known as superposition.
- A particle can spontaneously appear out of nothingness and disappear back into nothingness, an extraordinary fact indeed.

- A radioactively decaying nucleus of an atom can spontaneously break down into other nuclei without any known reason or cause.

- Entanglement refers to the fact that once particles interact, they become entangled, regardless of the distance between them. They are forever entangled. They are connected by unseen energy, information that permeates everything in the universe.

- Entanglement is only possible because everything in the universe is connected at some level. Everything is energy-information connected at the subatomic level. Everything in the universe contains information about everything else.

- One of the applications of entanglement is teleportation. Scientists have been able to teleport particles from one place to another.

- Particles can cross from one place to another without any restriction. This is called quantum tunnelling.

Quantum Field Theory

In an article entitled "There Are No Particles, There Are Only Fields," Prof. Art Hobson asks: "Are the fundamental constituents [of the universe] fields or particles?" He responds by saying: "As this paper shows, experiment and theory imply unbounded fields, not bounded particles, are fundamental" (2013, 1). This statement sums up what the proponents of QFT strongly believe are the fundamental components of matter. What is QFT? It is a theoretical framework that combines classical field theory (a physical theory that predicts how one or more physical fields interact with matter through field equations), special relativity (an approach that deals with conditions in which gravitational forces are not present), and QM but excludes general relativity's description of gravity. In QFT, particles are treated as excited states of their underlying fields. The fundamental objects of QFT are quantum fields.

The idea behind QFT is that what we call subatomic particles (electrons, photons, quarks, etc.) are simply excitations in the fundamental fields. Many physicists have shifted to an all-fields perspective, and suggested that there is enough evidence for physicists to agree on an all-fields view. See, for instance, the work of Prof. David Tong (2009) and Dr. Ethan Siegel (2018) for in-depth review. According to QFT, the universe's fundamental building blocks are not discrete particles, such as electrons, but are continuous fluid-like substances called fields. Examples of these fields include the electric field and the magnetic field. An image of a magnetic field emerging from a magnet is shown in Figure 1.2.

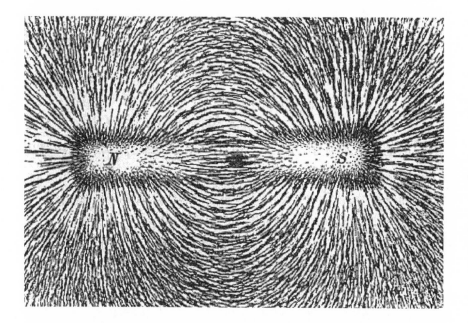

Figure 1.2 *Fields emerging from a magnet. Image by Newton Henry Black via https:// commons.wikimedia.org/w/index.php?curid=73846 (Creative Commons CC-BY-SA-4.0 license)*

Instead of talking about particles, those working in the QFT field discuss everything in terms of fields. That is why they talk about the electron field, quark field, gluon field, and Higgs boson field. A particle is simply a tiny ripple of an underlying field, shaped into a particle by the effects of QM.

We learn from QFT that there are 12 fields. These fields fill our planet and, indeed, the entire universe. We know from QFT that the 12 fields produce matter and four other fields, i.e., the forces. In a nutshell, our universe is the blending of these 16 fields interacting together. In addition, we have so-called Higgs fields. These are responsible for the mass of everything. The 12 matter fields are made of six quarks (up/down, charm/strange, and top/ bottom) and six leptons (electron, muon, and tau and their respective neutrinos).

On the other hand, the four force fields embodied by the electromagnetic force, carried by photons, create electric and magnetic fields. They are responsible for chemical bonding and electromagnetic waves. Several researchers have discussed these forces. For instance, QFT's fundamental concepts and theory have been discussed by Art Robson (2017).

I have introduced the idea of fields to explain the nature of matter from the viewpoint of a sub-branch of physics called QFT. From this viewpoint, electrons in atoms are not particles or solid stuff; they are probability fields. The physical world is determined by the interference (computation) of these fields (waves). My understanding is that the calculations of waves create the structure of biological matter that we see in the physical world. These interactions are continuously ongoing, running the universe. These fields behave like an intelligent system.

These fields are embedded with information and perform complex computations; complex mathematical operations are taking place, the fields behaving like intelligent living entities that know what they are doing. The outcome of any of their interactions is already known; during the ongoing computation and information exchange, reality becomes nonmaterial. One way of putting it is illustrated in the work of Diogo Valadas Ponte and Lothar Schäfer (2013, 605): "The forms are real, even though they are invisible because they have the potential to appear in the empirical world and act in it. We must now think that the entire visible world is an emanation out of a non-empirical cosmic background, which is the primary reality, while the emanated world is secondary."

To stress once again, QFT tells us that matter (i.e., the universe) is not made of particles but rather of fluid-like substances known as fields. These fields are spread throughout the universe. QFT extends the notion of the particle from QM from a single particle to fields that extend to the universe. The premise of QFT is that matter is made of fields. QFT informs us that atoms are made up of vortices of energy. All particles in the universe, including those in you and me, are waves of the same underlying field. The particles in an atom are simply ripples of the same field as the particles in a river.

I am trying to understand QFT theory in terms of information and consciousness. What the proponents of QFT call fields are an intelligent system (intelligent fields). The fields carry information, perform computation, and self-organize. This process hints at the presence of intelligence or consciousness. I genuinely think that there is a need to develop a sub-branch called the computational theory of fields.

To sum up, my understanding of QFT, combined with the concept of information, tells me that matter in its true and original form is consciousness. Matter is simply consciousness that materializes and can be seen through our senses of perception. Matter can be considered as a passing state of consciousness. It is through movement, computation, and information exchange that consciousness turns into matter.

QFT is a theory of intelligent fields, meaning of information, and consciousness. The fields carry information that performs computation, self-organizes, and thus shows the presence of intelligence. Information exchange and interactions between fields give rise to what we see in the physical form or object. The computational theory of fields may provide us with a detailed picture of what the universe is computing and what kind of operations are being performed.

String Theory

This is the most promising candidate for a ToE. It is important to stress that ST teaches us that matter is made of particles. These subatomic particles are not point-like objects, such as the one postulated in the SM, but are relatively tiny strings. Figure 1.3 shows tiny vibrating strings. String theory replaces subatomic particles with strings that are either closed or open loops. These strings vibrate at different frequencies, and each different vibration gives birth to another particle.

String Theory has made little progress in the ongoing effort to find a ToE. There are still many phenomena in the universe that cannot be explained in terms of our leading theory. And there are also many other phenomena that we do not know of and cannot explain, perceive, or detect through our understanding of ST. It appears that a single approach, such as ST, is too limited. One of the criticisms of ST is that the reality of consciousness has not been included in any idea of the universe or ToE. Despite its success and achievements, many physicists have realized that ST cannot be a ToE due to its shortcomings.

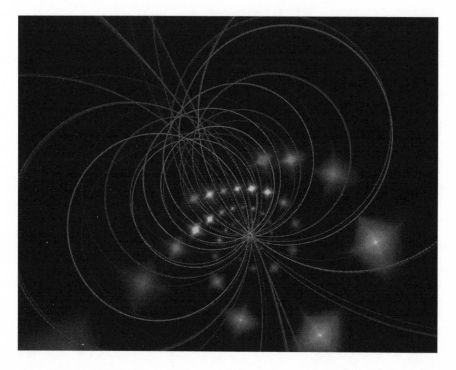

Figure 1.3   *In string theory, all matter, force, and particles are replaced by tiny vibrating strings*

Information, Computation, Self-Organization, and Consciousness

Our eyes see matter as solid, hard, and concrete, simply because our eyes are meant to see it that way to make sense of the world. The fundamental entities of matter are not atoms, electrons, quarks, or fields but information and consciousness. Matter is not an illusion, but it is not physical. The physical world is not physical at all. It is made of energy and information and is imbued with consciousness. Our eyes are meant to see the material to help us make sense of the world, but we are not physical at the fundamental level. We can only hear and see on a certain frequency.

Atoms and subatomic particles register information, which is energy vibrating at a certain frequency; this particle-energy is imbued with information and consciousness. Atoms and subatomic particles process, calculate, analyze, deduct, compute, create, and manifest themselves in different forms in the universe. They are intelligent information processing systems. The connection between matter and consciousness shows that what we call matter is not a physical object but a conglomerate of energy, information, or consciousness. Matter itself has some rudimentary form of consciousness.

This study is telling us something about matter—that it is not a physical object. By digging deeper, the conscious nature of matter is uncovered. Looking closely at matter, energy, information, and consciousness, we can see how they are interrelated. Matter is energy; energy behaves like an intelligent system, which is simply consciousness. The latter is primary. We tend to ignore the link that binds us all together; we forget the concept of wholeness, that we are one. We break everything into parts, embracing the reductionist approach in our vision of the world. Thus, it becomes incomplete. Our idea of the universe is not accurate. Our eyes are deceiving us into thinking that we are only physical beings.

The universe is real but depends on consciousness for its existence. It is permeated and imbued with consciousness. Without consciousness, there is no life. Our understanding of the structure of matter teaches us the intrinsic properties of the subatomic particles that compose it. We learn that so-called matter is inseparable from consciousness. It is a manifestation of consciousness. A more in-depth study of atoms reveals a world where we realize that what we think of only as a solid is imbued with consciousness and intelligence. The physical world and consciousness are thus one, and this insight gives us a better description of the universe.

It is clear that matter is not solid and that it changes its form. What we see as a solid object is only a perception. Perception is the key; a human being can only perceive and hear what is in a specific frequency range. What we perceive as concrete is not necessarily solid. We have learned that supposedly solid matter is not solid. The further we dig into matter, the more evidence we find that intelligent processes are at work at the most fundamental level. Matter as we experience it seems to be the result of human consciousness interfacing with the so-called constituents of matter (electrons, nucleus, quarks, etc.), which are waves of energy, i.e., information. Thus, the object I see in front of me is simply information-energy received by my brain and translated into a picture called an object.

To sum up this chapter, I would like to say that matter comes from consciousness. Matter in its true and original form is consciousness. Matter is simply consciousness that materializes and can be seen through our senses of perception. Matter can be considered as a passing state of consciousness. I have shown that through movement, computation, and information exchange, consciousness turns into matter. The implications for humanity are profound: the universe is made of information, is immaterial, mental, and spiritual, and consciousness is fundamental.

A consensus may emerge that through the understanding of digital physics (the physics of information, computation, self-organization, and consciousness), we may have the best alternative theory around. Digital physics offers us a consensus among the myriad quantum theory interpretations. We have realized that the main ingredients of physics, chemistry, biology, and psychology are information and consciousness not particles or fields.

PART TWO

The Illusion Of Energy

The Origin of Energy

Today the vacuum [of space] is not regarded as empty. It is a sea of dynamic energy, like the spray of foam near a turbulent waterfall.

—HAROLD E. PUTHOFF

From Chapter 1, we have learned that matter is not a physical object and that the particles that make up matter are simply vortices of energy, as illustrated in Figure 2.1. Atoms consist of energy or vibrating energy fields, such as the electron field proposed by the QFT proponents. For the proponents of QFT, particles are simply fields, so they refer to the atom components as electron field, quark field, gluon field, or Higgs boson field. These vortices of energies or fields carry information, compute (calculate), and behave like intelligent systems. For instance, an electron is imbued with properties and information, such as its electric charge, atomic charge, mass, and atomic mass. And when two electrons combine, they exchange information; similarly, when two atoms combine (compute), they create a molecule by exchanging information.

Figure 2.1   *An illustration of the hypothetical vortex*

We learn that what we think of as a solid is only an appearance caused by limitations in our perception. An atom is made of empty space, with energy popping out and in. Solid matter is not a physical object. At its most fundamental level, I have shown that matter does not consist of so-called particles. Matter, energy, or fields—whatever we want to call it—is embedded with information and is continually computing. Matter behaves like an intelligent entity, a form of consciousness. Now we should ask ourselves, what is the origin of this energy? What is the source of the different types of energy found in the universe? These are difficult questions to answer, but we should tackle them head-on in the quest to find the true nature of energy.

The nature of energy remains a mystery. Many writers have pointed this out. It is well documented that many in the scientific community refer to energy as an abstract property. Nobody has ever seen energy, only its manifestations. Various notable writers have questioned it. For instance, Abbot and Van Ness (1972), Rose (1986), and Pepperell (2018) have gone so far as to argue that "Energy is a mathematical abstraction that has no existence apart from its functional relationship to other variables" (2018, 4). It should also be pointed out that there are two primary forms of energy: kinetic and potential. Kinetic energy is possessed by objects due to their motion, while objects have potential energy due to their relative position or configuration. All other energy types, such as electric, solar, thermal, chemical, and atomic, are in themselves forms of either kinetic or potential energy.

Although several questions are being asked about the nature and origin of energy, possible answers may lie in an atom's composition. I have already pointed out that an atom is 99.99% empty space and that the space between the nucleus and the electrons is empty. This so-called empty space is filled with energy. It may consist of energy, known in QM as ZPE. QM predicts the existence of ZPE for the strong, the weak, and the electromagnetic interactions, and it refers to the lowest quantized energy level of a quantum mechanical system or a system's energy at temperature zero kelvin or −273 degrees Celsius.

We learn from physics that at zero kelvin, the particles of matter stop moving, and all disorder disappears. But is this true? No, simply because energy remains when all other energy is removed from a system. A typical and well-known example is the case of helium. For instance, if the temperature of helium is lowered to absolute zero, it remains a liquid. It does not freeze. This is due to the presence and the irremovable ZPE of its atomic motions. It is only by increasing the pressure to 25 atmospheres that it freezes. Is this ZPE the origin of the energy that binds everything together in our universe? Or is this the energy that gives so-called matter its solidity?

Let me dig deeper by looking at the theoretical and experimental works done in QM since the early 1930s by those keen to discover the origin and prove the existence of the ZPE. Many researchers suspect that the ZPE may be the source of so-called atoms' energy, although its origin remains a mystery. These discussions have only one aim, to discover whether the work done over several years and the contributions made by various authors have lain the groundwork for understanding the origin of energy or the transformation of energy to matter (or vice versa).

Let me start clearly by stating that the ZPE is one of QM's predictions —i.e., that each cubic centimeter of empty space or vacuum contains an amount of energy. In everyday language, a vacuum simply means empty space. There is nothing. However, its meaning in physics is not the same as its general use. A void is not empty at all. It contains fleeting electromagnetic waves and the so-called virtual particles. Alternatively, we can understand the concept of the ZPE from the well-known process in QM called quantum emission/absorption. This process of exchanging quantum information is crucial to comprehend ZPE, which may offer significant insights about the nature of life, the universe, and beyond. And it brings the role of information in physics to a fundamental level. As a result, the phenomenon of emission and absorption of energy by all physical objects at the quantum level carries information about those physical objects' entire history.

Several works by the distinguished physicist Harold Puthoff (1988, 1994) suggest that the ZPE is the primary energy, the source that connects everything in the universe. This sea of energy seems to be fundamental, the source of all energy. It penetrates everything in the universe. In *Digital Physics: The Universe Computes* (2017, 28), I elaborate on this concept by stressing that one of quantum theory's (QT's) predictions is that empty space or a vacuum contains residual background energy called ZPE. Quantum theory predicts that all of space must be filled with electromagnetic zero-point fluctuations or zero-point fields, thus creating a huge, universal sea of ZPE. Why ZPE? Experiments have demonstrated beyond any doubt that at temperatures of absolute zero (−273 degrees Celsius), elementary particles continue to exhibit energetic behavior.

One of the earliest researchers to study this phenomenon was Max Planck in 1911. The concept was later taken by Albert Einstein and Otto Stern, who wrote a paper to prove the existence of ZPE. Unfortunately, in the same year, they changed their minds about the ZPE after further work by Max Planck (his second theory showed that the ZPE may not apply to their example). Several other physicists and chemists, such as Paul Ehrenfest and Peter Debye, developed and extended this idea. The central concept of ZPE is that so-called empty space is overflowing with energy. Any empty space between atoms, including the distances between planets, is filled with ZPE.

Further work on the concept of ZPE was illustrated by Hendrik Casimir and Dirk Polder (1948), who proposed the existence of a force in vacuums, following the experimental work that showed that two mirrors placed close to each other were attracted (see Figure 2.2). This force of attraction came to be known as the Casimir effect. This work showed a force existing in a vacuum. This source of energy is called the ZPE. Therefore, QM teaches us that this ZPE is something that we cannot see with our naked eyes but that is there and can be experimented with.

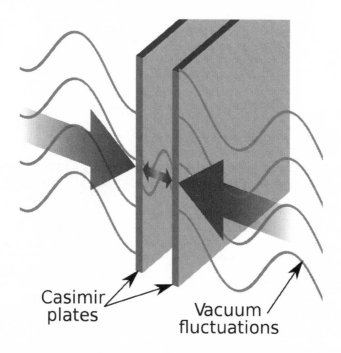

Casimir plates

Vacuum fluctuations

Figure 2.2 *The Casimir effect*

Understanding the attraction between two surfaces in a vacuum has led to a better understanding of the concept of energy. The phenomenon is known as the Casimir effect, while the force between the mirrors is called the Casimir force. Casimir remarked that the two mirrors facing each other in a vacuum were mutually attracted (every field carries energy). Although it was difficult to measure the so-called Casimir force, one of the first attempts was conducted in 1958 by Marcus Sparnaay at Philips in Holland. He went on to show that the outcome of his experiment agreed with Casimir's theoretical prediction.

Further work was done by Steve Lamoreaux, who measured the Casimir force between a 4 cm diameter spherical lens and an optical quartz plate about 2.5 cm across. The two plates were coated with copper and gold. Lamoreaux accomplished this experimental confirmation of the Casimir effect's existence in 1997. He demonstrated the Casimir force with an accuracy of 5%. Commenting on this experiment, Astrid Lambrecht (2002) remarked, "When Lamoreaux brought the lens and plate together to within several microns of each other, the Casimir force pulled the two objects together and caused the pendulum to twist. He found that his experimental measurements agreed with theory to an accuracy of 5%." Other experiments have been performed to prove this effect and force; see, for instance, Umar Mohideen and co-workers' work.

Although the ZPE or vacuum energy continues to be a subject of debate and continues to generate debate and passionate discussion in academia, it has been proven to exist. It does not matter whether we call it the ZPE, vacuum energy, or the Casimir effect; it has been proven experimentally to exist. The ZPE is ever present in a vacuum. What we call an empty void is simply full of energy. The energy is present even at absolute zero temperature, even when there is no matter. The ZPE registers information, such as binary digits and qubits. It processes, computes, and is the carrier of information in electromagnetic waves and other forms of transmission. It is the carrier of information and connects everything to everything else in the universe. It is the ongoing processes, computations, and information exchanges within that give us matter. The ZPE seems to be the total primary energy, the total intelligent energy, which behaves like "consciousness."

Energy and Consciousness

Energy and information are two different things. Energy is a property of matter, and we have learned that both are the same thing. Energy can be transferred from one object or system to another and converted from one form to another. However, it cannot be created or destroyed. Energy is needed, for instance, to transfer information from one place to another. The information takes different shapes, such as image, voice, and graphic, and can be transmitted by radio, the internet, satellite, or TV.

On the other hand, there are different types of information, depending on the field. In biology, with biological systems, such as cells, organs, organisms, and DNA, genetic information is transmitted across generations. When two atoms collide in physics, they exchange information and register bits of information (such as speed, temperature, position, or velocity). Edwin Parker (1974) argued that "Information is the pattern of organization of matter and energy."

Combining what we have learned about the ZPE so far and what I have discussed about consciousness, we are slowly, slowly beginning to understand how we are all connected through the ZPE. The connection between the ZPE and information is clear. Information is everywhere in the universe. But that thing, the force that brings everything together, is consciousness. Consciousness brings information and energy together. Consciousness is thus primary. Everything is made of energy and information, but without consciousness, nothing can move. What goes on is simple: Everything in the universe consists of energy, or ZPE, like vortices and information. The unseen, ongoing computation creates reality or the physical object. The analysis is continuously ongoing; the calculation is made instantly and precisely. But when we look at something, we only see the object's image not the object itself; because this computation is ongoing, our eyes are meant to see the physical object not the constant movement.

Newtonian physics is so deeply ingrained that we are genuinely convinced that matter is solid, and it is difficult to change that view. But we must always remember that the notion of an atom surrounded by particles is simply a model. We see matter as a physical object because our eyes are meant to see it that way to make sense of the world; otherwise, if we saw everything as energy or information, we would not be able to make sense of the world in which we live. It would be chaos. It is clear now that our eyes perceive things as solid so that we can interact with the so-called physical world. In reality, we do not see the ongoing process, the invisible computation continuously taking place.

As I have just explained, we are convinced that matter is solid simply because we touch or hold it in our hands. This approach is misleading and misguided. Those who think that matter, atoms, or particles are solid objects are misled. Energy is a fundamental entity of our physical world. Matter and energy are different manifestations of the same thing. Energy appears to be the product of consciousness. Quantum Mechanics, through the double-slit experiment, shows that consciousness can alter the nature of matter and energy, positioning itself as primary. For some researchers, the sea of energy (ZPE) is simply the driving force, a form of consciousness. Through this discussion, it is becoming clear that we should take the meaning and implications of these discoveries seriously.

We now understand that everything supposedly made of matter is energy. Matter, atoms, a tree, a table, a computer, it is all part of an interconnected web of electromagnetic waves. Electromagnetic (EM) radiation carries energy and information. Each wave has a specific direction, frequency, and polarization state. This web of energy found in each entity of the universe is known as consciousness. Each object emits, receives, and stores information and energy. All objects have an energy field with a vibrational frequency. Vibration refers to the oscillating and vibrating movement of atoms and particles caused by energy. Frequency is simply the rate at which vibrations and oscillations occur.

We are convinced that matter is solid simply because we touch or hold it in our hands. This is a misleading and limited view. Let me make it clear: Matter, atoms, or particles are not solid objects. Energy is a fundamental entity of our physical world. Matter and energy are different manifestations of the same thing.

Implications

Quantum Mechanics is shedding light on the true nature of reality and challenging the generally accepted view of matter and nature. Although experimental verification has not contradicted it, the conclusions of many of its experiments are disturbing, and the idea that its extension, digital physics, may offer an accurate description of nature is met with cynicism. The ZPE connects everything that carries information. Matter is energy. Energy carries information. All this points to the role of computation in the universe and the fundamental role of consciousness.

The implications of these discoveries are massive, even though not many people think about them in the way I have discussed it in this book, even among physicists who are aware of the facts. Understanding this brings a new perspective for understanding ourselves and the world in which we live and gives us a bigger picture of the universe. This knowledge is useful and should benefit all of us. Our world is not limited to what we can see. In a nutshell, there is more than we can perceive with our senses.

The implications are clear:

- Each one of us is made of vortices of energy that are constantly spinning and vibrating, each one radiating its own frequency (emitting, absorbing, and computing information).

- It is clear that the universe is not some sort of physical, material object but rather is immaterial (energy, waves, and information).

- Energy is constantly changing its form and vibrational frequency and transporting information. This energy travels in waves and carries

multiple pieces of information. We receive and processes this information transported by these electromagnetic waves.

- This concept of energy–information--consciousness should be extended
in all areas of academia.

- We are all connected by this ZPE; separation is an illusion. We should all
take our responsibility and behave accordingly.

- We should not do things that harm others; harming someone will ultimately harm ourselves and the whole of humanity because we are all
connected.

- Our physical body is not the only thing we ought to be aware of and
take care of. We are all energy-information, conscious beings. We are
continually emitting and receiving.

- We should apply the concepts of energy, information, and consciousness
in our lives so that they benefit ourselves and all humanity.

- The evidence supports the view that matter is made of energy (matter
registers information, processes it, and presents us with what we see).
Matter has elementary consciousness. I would like to extend this to
human beings or everything in the universe. We are all emitters and
receivers because we are all made of the same fundamental entity.

- The role of consciousness in shaping the nature of physical reality is so
important. Thoughts, emotions, and feelings result in different electromagnetic frequencies. And this has an impact on the physical world. For
instance, when an atom gains or loses an electron, it changes its state; it
is no longer neutral. If an atom gains or loses an electron, it becomes an
ion: If it gains a negative electron, it becomes a negative ion; if it loses
an electron, it becomes a positive ion.

- Our emotions and thoughts and the vibrations of our bodies attract
things like a magnet into our consciousness that are vibrating at the
same frequency as ourselves.

- Energy is the building block of matter. The same energy you are made
of is also the one that makes up plants, trees, animals, and minerals. We
are thus connected, entangled by the energy we are all made of; that is

why our perceptions, vibrations, feelings, and thoughts (both good and bad) affect others. Positive thoughts influence those around us. Similarly, negative energy and emotions affect those within our perimeter.

- Our thoughts are, in a way, waves filled with energy. Remember that we are part and parcel of this "universal consciousness." We are who we are; our planet's state is the result of our collective thoughts, and emotions. We must and should control our actions and feelings because these affect everything else in the universe.

- Material bodies vibrate at higher or lower frequencies. They then emit and receive. For instance, an exchange between a human being and something with a high vibration affects them. The cold (lower vibration or frequency) and heat (higher vibration) likewise affect our own behavior.

- The concept of energy leads me to say that thoughts, emotions, and feelings are forms of formless energy released by bodies.

Before summarizing this section, let me remind my readers that we learned that the smallest form of matter is particles. Particles form an atom; in turn, when atoms combine to form molecules, these are what we see as matter. Now, I have already said that particles are simply vortices of energy, leading to the conclusion that everything is energy imbued with information. Matter becomes then a gigantic concentration of energy. Quantum Mechanics tells us that matter is the same as energy. Matter is energy that has a solid form, energy that we can touch and see.

On the other hand, energy is potential matter that has yet to solidify or form. Quantum Mechanics has discovered through experiments and observations facts that we could never have thought possible. Many people are still in denial. This realization has improved our understanding of reality. We can no longer think of the universe as only a physical entity, like concrete, in which the things we see and touch are all that exist. The universe is embedded with information. Information is embedded in the physical constructs of the universe and is transmitted everywhere.

I would like to end Chapter 2 with a quote from Albert Einstein: "Everything is energy, and that's all there is to it. This is not philosophy. This is physics."

But I would like to add that energy carries all types of information and that the universe computes and is a self-computing consciousness. Although energy is a form of consciousness and a medium, it is the information it carries and the ongoing computations that appear far more important. The entire process is guided by consciousness.

PART THREE

———

Revising Our Concepts of Time, Space, and Perception

Time Perception

Time is change, nothing more, nothing less. We do not see things as they are but as the brain interprets them for us.

—JULIAN BARBOUR

Understanding the nature of time and its place in physics appears to be one of the stumbling blocks for unifying general relativity and QM. On the one hand, a sizeable number of physicists claim that time does not exist; on the other hand, we hear the argument that time is real. Therefore, in this chapter, I would like to answer several questions about time and then explain it in terms of consciousness. Is time real? Is it an arbitrary measurement? Is time something we have created to make sense of the world? Is it only a concept that exists in our minds?

To answer these questions, I would like to discuss and frame the concept of time by surveying and looking at various viewpoints of physicists, philosophers, and researchers from six perspectives: (1) our everyday use of it, (2) the views of two physicists, (3) classical mechanics, (4) relativity, (5) quantum theory, and (6) consciousness. I am aware that what we call time is something we cannot see or touch. It is an invisible entity, something that is not physical. Generally, when dealing with something we cannot see or touch, we have to be very careful in answering these types of questions.

By examining these different views of time, it is possible to answer these questions, and we will learn more about time than we presently know. Also, we might learn a little bit more about what we do not yet know and about why, despite several centuries of research, physicists have not, so far, been able to come up with a single consensus about time. The problem of time is intrinsically related to the nature and perception of matter, energy, and consciousness and cannot be separated from them. Time should, indeed, be understood by looking through the same lens as the intrinsic nature of matter and energy, as elaborated in the previous pages. It is along these same lines that I would like to examine the meaning of so-called time.

In the physics community, opposing camps have developed their own perspectives of time. Some lean toward the view that time is an illusion—that it does not exist and has no direction. At the same time, others firmly believe that time exists and moves in one direction only. My aim here is to clarify the concept of time by relating it to matter, energy, and consciousness. Because they are related, to decode the meaning of time and show the role that consciousness plays in understanding time is crucial. By doing this, I posit the function of consciousness as fundamental.

The concept of time is found in all theories of physics. I hope that the approach I have adopted in this book, using the images of matter, space, and energy, is persuasive. The reason is that I am convinced that the idea of time cannot be understood independently of matter, energy, perception, and consciousness. I posit that consciousness is the key, the final pass in understanding everything, even time. Recently there have been many debates concerning the nature of time. For example, a book published by a renowned Italian scientist, Prof. Carlo Rovelli, suggests that time does not exist and is simply a construct of our mind. Furthermore, a conference was organized in 2016 by Lee Smolin and colleagues in Canada, aiming to learn more about time and its meaning. These two recent developments have led to various discussions on the internet and various social media platforms.

In 2016, Lee Smolin, Marina Cortes, Roberto Mangabeira Unger, and other colleagues organized the Time in Cosmology Conference, which was attended by over 60 physicists along with philosophers and researchers from other fields of science. Over a period of four days, these scholars tackled several questions. For instance, how do we reconcile our perception of the passage of time with the idea of a static, timeless universe? During the meeting, various models to explain time were presented. For instance, George Ellis spoke and discussed his "evolving block universe." This is the idea that space–time is best represented as an evolving block universe (EBU), in contrast to the block universe view. For a detailed discussion of the conference, see Falk (2016).

Several years ago, various solutions to the problem of time had been put forward. The dominant view among physicists is that time poses a great challenge in the unification of general relativity and QM. That is why over 40 years ago, two eminent physicists, John Wheeler and Bryce DeWitt, collaborated and came up with an equation that provides a possible road map for unifying relativity and QM. In the Wheeler–DeWitt equation, time disappears. Although up to now, nobody has succeeded in using this equation to combine QM and general relativity; the quest is ongoing.

Meanwhile, Don Page and William Wootters' experiments in 1983 are well documented. They came up with the proposition that quantum entanglement may be a way to resolve the problem of time. They showed that a clock entangled with the remainder of the universe would seem to tick when viewed by an observer within that universe. In 2013, a group of physicists demonstrated this effect in a physical system, using a model of a universe that contained two photons (Moreva et al. 2013). This was a step forward in showing that in practice, this phenomenon can occur. It is evident that many solutions have been put forward, but so far, none of them is satisfactory.

Everyday Meaning of Time

In physics, we use a framework known as space and time or space–time, which is where events take place. The events occur in a location (space) and in a moment (time). For instance, the Dutch theoretical physicist Gerald 't Hooft defined time "as the order in which our models for nature predict, prescribe or explain events" (2018, 1). This definition is based on the need to construct models to explain our universe, argues 't Hooft. He went on to stress that physics needs a description of an ordered time coordinate.

Another interesting definition of time is from the article titled "Time" in the Internet Encyclopedia of Philosophy written by Bradley Dowden: "Time is what a clock is used to measure. Information about time tells the durations of events, and when they occur, and which events happen before which others." Time, whatever we do daily, is one of our references. When we wake up in the morning, we look at our clock and plan our journey accordingly. This time gives us the impression that time flows in one direction; it provides us with a sense of the present, past, and future.

Generally, a clock is used to measure time or read time. This is the generally accepted definition of time. A watch is a device that reads time. That is why, in our modern world, we have divided time into second, minute, hour, day, week, month, and year—solely to make sense of the world, to keep a record. But the truth is that what we measure is not time but memories or documents that we maintain. What we call time is simply the span that separates events. Time is viewed as a flowing arrow, the one arrow of time or the one-way direction. This concept of time as moving in one direction is sometimes referred to as the "asymmetry of time." In this view, time is something irreversible, as the passage of one season to the other. The sun rises in the east and then sets in the west: this is the ordinary view of time.

The Internet Encyclopedia of Philosophy tells us that time is about measurement and that this measurement is done during a period in which something happens. But the truth is that from this definition, we do not learn what time is. The question, once again, is, what is time? What is its nature? Is time a real phenomenon? Why do we have the impression that it flows only in one direction when there is not a single physical law of physics that forbids time reversal? Several scientists have attempted to answer these questions. Unfortunately, we have more questions than answers in the quest to find the truth about the intrinsic nature of time.

Questions continue to be raised about what time is. Many suspect that what we call time is only an artifact of our perception and that time may not be a fundamental property of the universe. There are those who question whether our idea that time flows in one direction is misleading. We have the feeling that time flows in one direction. It is clear now that from our daily experience, our concept of time is linked with our day-to-day actions—our work, and daily chores. In this regard, our sense of time is closely related to a division of labor, activities, and recording of events. It then follows that time appears as a construct of our brain to make sense of the world and the change. In reality, the present, past, and future are all happening now— right now. We must differentiate between changes taking place and time itself. Our brain has a way of perceiving changes taking place. And it is that ongoing change, wrongly or rightly, that we are referring to as time. In this sense, time feels real for us.

The View of Two Physicists

PROF. CARLO ROVELLI

In his book *The Order of Time*, the respected theoretical physicist Prof. Carlo Rovelli, one of the leading researchers in LQG, posits that time is an illusion and that we should revise our understanding of it. He explains that the concept of time in both classical mechanics and Einstein's theory of relativity is a simplification. He uses the notion of "events" to helps readers understand physics in a new way. Events, according to Prof. Rovelli, are the fundamental constituents of the world. They encompass location and time, i.e., when and where something takes place. An event is represented by a point in space–time (a point in space at a particular moment in time). He argues that we can understand physics in a better way by looking at these events.

As you read this book and some of Rovelli's articles about time, one thing becomes clear: time does not exist. He postulates that there is no time variable in the fundamental equations that describe the world. He stresses that what we call time is simply chronology, memories, and stories—processes in our brains to make sense of our existence. Furthermore, he does not mention space or time; there are only processes or changes that transform physical quantities from one to another. Prof. Rovelli declares that we see orderly, chronological events, such as the past to the present, because we have superimposed order on everything we are doing. We are putting everything we do in a linear order to make sense of the world, which gives us the concept and perception of time.

Our everyday experience of time and perception tells us that time flows in one direction. It flows like water in a river. However, Prof. Rovelli is telling us that there is no such thing as past or future and that time does not exist.

He expresses his ideas clearly and suggests giving up the concept of time in physics. His opinion and view on the issue of time are clear; he is genuinely convinced that the fundamental description of the universe must be timeless. My reading of Rovelli's work suggests that our brains process and create time perception. What is going on is that because we cannot see the ongoing computation taking place everywhere in a timeless and dimensionless universe but see instead a fixed "event" as a stable, material, physical entity, we think of the event as real. However, the computation is continuously taking place.

DR. JULIAN BARBOUR

Dr. Barbour has been studying time for over 60 years. He has gone further than any of his compatriots and asked more profound fundamental questions to uncover time's truth and meaning. He has looked at many aspects of time—duration, a clock, motion, and space; the purpose of time itself; and whether today is the same as yesterday. He stresses how this notion of time has been neglected—even Albert Einstein questioned only certain specific aspects of time. He hopes that tackling these questions that have been neglected over the years will uncover some elements of so-called time that we ignore.

Several years ago, in the late 1990s, Dr. Julian Barbour set himself the task of putting forward a radical and thought-provoking concept of time. In his book *The End of Time: The Next Revolution in Our Understanding of the Universe*, and in several of his articles, he presents several ingenious and original ideas and has a genuinely new vision. He argues that the concept of time moving in one direction is an illusion. He stresses that we inhabit a universe with no past or future and live in an internal presentation. By questioning time, he interrogates whether the notion of time is fundamental at all.

One thing that strikes me the most is his attempt to build up a theory "of relationships between things" in contrast to most theorists before him, who have generally come up with abstract methods disconnected from reality or ill-defined, complex notions of space and time. He is right when he points out that the world in which we live is relational, i.e., that things relate to one another. And in this mutual relationship, the notion of interconnectedness is stressed. He posits and emphasizes that time, as such, is an illusion and does not exist. In an interview with John Brockman, he speaks instead of instants of time or "Now." For instance, he says, "There are many Nows, all different from each other. That's my ontology of the universe—there are Nows, nothing more, nothing less" (Brockman 1999).

He makes the interesting point that our belief in time is greatly influenced by classical mechanics, particularly by the concept of absolute space and time. As a solid, static entity, the idea of matter creates the false notion of a fixed reference from which the moving forward of time and change take place to become the norm. A common theme emerges from his work that is similar to one we find in the work of Prof. Rovelli. Both show that time, as an independent concept, does not belong in physics. Their idea of a timeless universe is worth exploring, and further research is needed to uncover the fundamental essence of time.

The concept of a timeless universe led me to reflect on the issue of time. Is there any clock in the universe in a physical sense? No, there is not; there is no clock ticking somewhere outside the universe. An ordinary clock or watch does not measure time at all. We have built equipment with rules to record the passing of days, weeks, months, and seasons. We have, in turn, divided time into seconds, hours, and minutes. In this sense, time is defined by the number of clicks or cycles. We do not see the time itself. Time is immaterial. So, in a sense, we are confusing changes; the ongoing computation that is taking place, we have decided to call this change time. So, in a way, it should not be called time.

Time is immaterial, unobservable; it is only a concept that we have intro-duced as a reference to make sense of the world, similar to how we view particles as physical objects. Particles are not solid, physical objects (as I explained in Chapter 1 and 2). This is misleading and does not give us an accurate picture of the world in which we live. What we call time may really be merely a succession of events and snapshots, changing constantly or steadily. The key here is *change*. Change becomes a synonym for time. Without that transformation, the notion of time would not exist. What we call time is simply change. Yes, of course, change is happening worldwide, where you are sitting, eating, and working. Change is taking place, alterations are being made—this dynamic is real and ongoing—but time is not. Time thus can be understood as a reflection of change.

Classical Mechanics

In classical mechanics, the notion of time is well documented. Sir Isaac Newton talked about absolute space and absolute time. The meaning of this is that regardless of whether something happens or not, time is passing; time passes uniformly whether or not something happens. In a deterministic theory, such as classical mechanics, time is a parameter that is used to describe a system or an object. The system–object path, time, is fixed.

Let me reiterate that time passes uniformly in classical mechanics, regardless of whether anything happens in the world. That is why we speak of absolute space and absolute time. Time is independent of human beings. This is how we generally think of time. Time is a parameter used to describe the path of an object or particle. The concept of symmetry is well documented in classical mechanics. A symmetry exists between the direction of time and distance, and one can always go back in the opposite direction—for instance, moving from a point x (t) and going in the opposite direction x (–t). This is to say that time is reversible in Newton's law of motion.

So, in classical mechanics, we talk of the reversal of the direction of time. As rightly pointed out by Görnitz (2014, 2), "The fundamental equations in classical physics are invariant with respect to a reversal of the direction of time. In the same way as a film can be shown backward, the physical models in classical physics can run forward or backward in time." The concept of an object or the path of a particle has a real meaning in classical mechanics.

Theory of Relativity

With the advent of the special theory of relativity in 1905, Albert Einstein revolutionized physics. He introduced the concept of space–time, which consists of three dimensions of space plus one dimension of time. The idea of space–time became ingrained in physics. An event takes place in space and time (the fourth dimension). Einstein's work has taught us to speak in terms of space–time. In 1916, he put forward the general theory of relativity, which combined space–time with the mass–energy equivalence. From the general theory of relativity, we learn that time is not absolute but relative to the observer.

Einstein realized that gravity is not a field or force but merely a space–time feature. Space–time is warped and curved mass and energy; this is what we refer to as gravity. With Einstein's special theory of relativity, we learn that time does not flow at a fixed rate. In general relativity, the concept of relative time is introduced. Larger and smaller curvatures affect how time is viewed. Time flows slower where the curvature is larger and faster where the curvature is smaller. The concept of curvature of space–time thus replaced gravity. General relativity tells us that time is not absolute but depends on the observer. Furthermore, in general relativity, time is relative and dynamic. Past, present, and future are not absolutes. Time in general relativity is not a universal constant.

The Block Universe Theory

According to the block universe, which underpins Einstein's relativity theory, time is taken as a four-dimensional space–time structure; time is like space. Each event or occurrence has its coordinates in space–time. Nothing happens in this picture; there is no change; past, present, and future are interwoven, as shown in Figure 3.1. In a nutshell, relativity's quintessence is summarized by the idea that everything is relative; there is no absolute time and no absolute space.

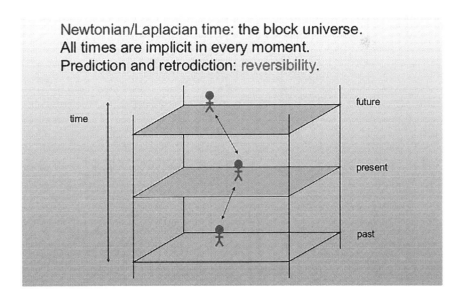

Figure 3.1 *The block universe. All times are absolute in every moment. Image courtesy of Sean Carroll via https://www.slideshare.net/seanmcarroll/setting-time-aright/3NewtonianLaplacian_time _the block _universe*

The block theory or the block cosmos is characterized by the idea that the universe is static and that past, present, and future all exist simultaneously and are real. Furthermore, it posits that the concept of the arrow of time or time flow is simply a mental construct. However, not all physicists agree with this idea of the block universe. Some physicists have stressed that the passage of time is physical. One notable critic of the block universe is Lee Smolin of the Perimeter Institute for Theoretical Physics in Canada. He says that "The future is not now real and there can be no definite facts of the matter about the future." What is real is "the process by which future events are generated out of present events" (quoted in Falk, 2016).

The block universe's proponents do not see time flowing as in the model of the arrow of time; nor do they see it as a sequence of unfolding events. Their idea is that both past and future are there as part of four-dimensional space–time. All times—past, present, and future—exist, despite our inability to experience past and future. Once again, I think that frequent change is what gives us the impression that time is moving in one direction, that it is flowing. Our brain interprets this ongoing movement and change as time, and thus we believe that time flows in one direction.

Quantum Mechanics

Although in QM the concept of time does not appear to be a fundamental element, nonetheless, it is there. Time is not something divided into a discrete entity or indivisible packets. In QM, not all properties of the universe are discrete. Electrons are discrete entities, like photons. Light, a type of electromagnetic radiation that travels in waves, is made up of quanta or photons. We would expect time to be discrete and divided into discrete quanta, but this is not the case. But we can learn a little bit about the possible discrete aspect of time from our leading theories of quantum gravity.

Although time is considered as smooth and continuous in physics, its quantization has been discussed elsewhere, particularly in LQG. Recent advances in both ST and LQG theory suggest that space and time are ultimately discrete. But most ST calculations take space–time to be continuous. On the other hand, in LQG, space and time are discrete—LQG quantizes space–time. For an in-depth review, see Smolin (2004).

According to the Big Bang model, which is supported by most physicists, time had a beginning 13.8 billion years ago. In QM, the concept of time, with its distinction between present and past, is not the same as in classical mechanics; the orderly idea of time is nonexistent. There is no order of events as there is in classical mechanics. From the QM viewpoint, in the microscopic world, the laws are time-reversal invariant. It is impossible to determine from the individual movements of particles whether they are moving forward or backward in time. At the quantum level, time does not exist; time is universal and absolute.

Quantum Mechanics, like other branches of physics, does not neglect time. However, the standard particle physics model is well documented, and it suggests that protons do not experience time. Some recent work in QM suggests that time does not exist at the quantum level, meaning that time is a function of relativity only.

The concept of the path taken by an object (particle) is meaningless, since there are only probabilities that a particle can be located at a given position—if a measurement is made at time t, for instance. A wave function describes a particle; the latter involves time according to a differential equation. But here, the process of measurement is crucial in understanding the concept of time because it introduces an asymmetry between the directions of time. In a way, we learn that the method of measurement divides time into three phases or periods: first, the time before measurement; second, the period during the measurement; and third, the time after the measurement.

Time becomes irreversible when we bring aboard the concept of the collapse of the wave function. As rightly explained by Calhoun137 in an article published on Medium (2014), "If a certain wave function describes a particle before a measurement, then the process of measuring the particle is said to 'collapse' the wave function into one of its fundamental quantum states, which has a definite measurable value. The collapse of a wave function is, in general, irreversible."

Further understanding of time can be learned from the idea of non-locality, which helps us uncover the truth behind time. Non-locality or entanglement is the transfer of instantaneous and simultaneous information through wave-like or field-like resonance regardless of the distance of separation. For instance, once two particles are entangled, they remain in communication regardless of the distance between them. If two entangled particles A and B are separated, they will always stay in contact, irrespective of distance. For instance, if particle A is moved, particle B does the same without delay. Why? This is simply because the concept of time flow or linear passage of time does not exist; time does not exist. The idea of time becomes meaningless.

We can learn much more about time by looking at various interpretations of QM. In the Copenhagen interpretation of QM, a particle's position is described by a wave function, which gives the probabilities of locating the particle at different locations. When the particle is observed, the wave function collapses; the particle is finally found in one place or the other. The observation affects the particle. The wave function contains all information associated with the particle, such as its initial position. From the Copenhagen interpretation of QM, we learn that the collapse of the wave function cannot be undone or reversed; the process is time irreversible. The Copenhagen interpretation is one of the leading interpretations of QM. It was devised between 1925 and 1927 by Niels Bohr and Werner Heisenberg. Aspects of this interpretation include the elimination of determinism and classical causality, the introduction of probability, and the consequent incomplete description of nature. One of the major points of the Copenhagen interpretation is that a particle can simultaneously be in two states.

One of the interpretations of QM is called the many worlds interpretation. In this interpretation, there is no one universe; rather, the universe is infinitely dimensional—many worlds exist side by side. This does not distinguish between a particle or a system before or after it has been observed. In this view, the observer is a quantum system that interacts with other quantum systems. So, when quantum systems combine or interact with one another, the wave function does not collapse but splits into alternative versions of reality. All information is conserved.

Quantum Field Theory

In QFT, we learn a lot about the concept of time due to antiparticle properties. An antiparticle has a negative charge. However, they are ordinary particles; they move backward through time. So, in QFT, the notion of particles moving back through time is a reality.

Paul Dirac's work embodies a little bit of the history of QFT's foundation and discovery in QM's early development. In 1928, he laid the foundation for QFT. He amazingly solved the problem of expressing QM in a form invariant under the Lorentz group of transformations of special relativity, leading to the introduction of his relativistic wave equation. Dirac's equation correctly described the electron's fine structure; moreover, based on his equation, he predicted in 1931 the antielectron or positron. His works led to him being awarded the Nobel Prize in Physics in 1933.

It is somewhat strange how Dirac predicted the existence of anti-particles. He was dissatisfied with the work of his predecessors, who attempted to combine QM's theories and Einstein's special relativity. To achieve this, Dirac used only the information he had about the electron, i.e., its charge and mass. With his mathematical skills, he arrived at a wave equation for a single electron. He was then able to predict the properties of the electron, including its spin and magnetic charges. This breakthrough not only unlocked the mysteries of magnetic and spin properties of electrons but had other implications as well. Dirac's equation had two solutions, one representing the electron, the other its opposite, a particle with a positive charge and negative energy.

This led him to conclude that each electron had an antiparticle. Dirac thus predicted the existence of antimatter. This was confirmed experimentally in 1932 by Carl Anderson. A few years later, antiprotons were produced for the first time and identified in 1955 by Emilio Segrè and Owen Chamberlain. A year later, in 1956, Bruce Cork discovered the anti-neutron at Lawrence Berkeley Laboratory.

Consciousness

In the previous chapters, I explained that what we call matter is energy embedded with information and a form of consciousness that moves like waves. During these movements and transformations, various operations are conducted, such as computation, that present us with what we see. Therefore, what we call matter is a form of intelligence, a state of consciousness. Extending this concept to human beings, consciousness, which is embedded in us, the human being, is primary. Consciousness is independent of the body. For instance, during a dream, our consciousness has access to the present, past, and future simultaneously. For consciousness, there are no such things as present, past, or future. Consciousness navigates in a multidimensional and timeless environment and has access to all information.

Consciousness is active both during sleep and when we are awake. Thus, we are endowed with not one but two recurring forms of consciousness: one we experience during the waking state and the other during sleep. Therefore, we can speak of "waking consciousness" and "dreaming consciousness," as rightly pointed out by Ullman (1999). Both forms of consciousness are found in human beings and other entities, such as animals, who experience them differently. That is why, during sleep, our sleeping consciousness can see, hear, and move. In a dream, there is no need for a physical body.

We can perceive without a brain and move around without our physical body. See, for instance, *Digital Physics: Decoding the Universe*, in which I cite many experimental works done over the years. Consciousness is a multidimensional, timeless entity that has access to present, past, and future at the same time. You see what is happening in the future during dreams because it has already taken place in the so-called present.

The perception of the flow of time results from our waking consciousness, from how our brains process information to help us move around. Our brain is only a receiver and processor of information. Whatever we see as physical is not material because during the processing of the information (or history) of every moment, the past, present, and future exist simultaneously. All timelines, everything about the present, past, and future, are interwoven. When we look, we only experience and see what is needed to create a recent experience during this ongoing computation.

But the truth is that the future is there in the ongoing computation. In a dream, we can move in three dimensions of time—so-called past, present, and future—at the same moment. It is like space and time do not exist. We navigate freely, without any obstacle, in these temporal and spatial dimensions. But during the day, we create our reality during this ongoing processing or computation. During this processing, our perception creates the experience of time moving in one direction, of the motion, unfolding, transformation, and action.

Time is an immaterial dimension; it is not the same as the dimension of space. As rightly pointed out by Ben Alvele (2020), "Time perception, therefore, differs from our other senses—sight, hearing, taste, smell, touch, even proprioception—since time cannot be directly perceived, and so must be 'reconstructed' in some way by the brain." The brain processes information, reorganizes it, and presents to us the physical form for us to understand.

The two previous chapters have discussed the intrinsic nature of matter, particles, and energy at length and concluded that they are not physical or material objects. We know now that time is also immaterial. These developments lead me to argue that the concept of time as we know it in the physical sense, measured by a clock, does not exist. Time is simply a memory of events. This leads to the fact that our everyday experience of time is misleading. It is merely something that our brain has arranged to make sense of the world, the so-called physical, material world. It follows that the so-called reality we experience is not an objective reality, as shown throughout this book.

In a way, during this computation, our brain or consciousness perceives the ongoing movement rapidly, giving us the impression that time flows. This rapid movement creates a sense of motion, direction, and time. The future is simply in this ongoing movement that we can see, and it manifests itself. This continuous processing creates a sense of continuity. We do not see the future direction because our brain is processing each moment in an ongoing computation. We are building our reality continuously.

During this ongoing computation, reality presents itself successively as the next moment, where each snapshot comes or manifests itself and then disappears before the next one is given to us. This is similar to what happens when one watches TV. The image of what we see comes in pixels that combine as snapshots succeeding one another, giving us the impression that what we see is a fixed image. Pixels are being added one by one to form the image in front of us, but we do not see this rapid ongoing computation; we only see the picture on our TV. Our brain processes this and experiences so-called reality in sequential order. These moments are continuously flowing and are not separate from one another.

Thus, we now understand how we experience our everyday physical world as a result of this ongoing processing or computation taking place. We only see part of it as physical, but the processing is continuously being done. We create the perception of time and space. What we call time is simply a flow of events. Our brains are perceiving this ongoing computation as a physical object, i.e., reality. In a sense, we learn that so-called space and time are not real. We have learned from the consciousness approach of time that all timelines simultaneously coexist during this ongoing computation.

There is no such thing as time flowing in one direction, as rightly put forward by Larry G. Maguire (2015):

The reality is there are no points that exist in the universe; therefore, there is no time. Time only becomes viable when we perceive there to be points of reference, and a distance travelled (by the observer) between those points of reference. Time becomes especially real when we perceive these points of reference to be life events. All things are of a fundamentally psychic nature.

Physical things are merely the collapse of psychic things into physical reality.

With the same reasoning, Gruber et al. stressed the fundamental role of consciousness in understanding the concept of time: "With increasing evidence that all perception is a discrete continuity provided by illusory perceptual completion, the lower-level FOT [(flow of time)] is essentially the result of perceptual completion" (2018, 125). Several other researchers have done unique work in the area of time consciousness. See Andersen and Grush (2018) for an in-depth analysis. Although there are many opinions about time, one thing is becoming clear, most researchers suspect that time may not be fundamental, even though our daily experiences suggest that time flows in one direction.

Some physicists, such as Carlo Rovelli and Julian Barbour, have argued that time is not real. DeWitt and Wheeler proposed to remove time from physics, as they saw it as a stumbling block in the quest for the QG theory and the ToE. Although they may have been right in their approach, my analysis suggests that they neglected the role of consciousness in physics, particularly in QM. The exception is Dr. Barbour, who referred to consciousness in his quest to decode time.

I believe that a theory of QG and a unified theory or the ToE are only possible if we incorporate information and consciousness in physics. As long as physicists continue to neglect the intrinsic nature of their studies' objects, such as matter, energy, particles, and time, we are unlikely to make much progress. And with the current direction taken by most physicists who support ST, I do not think it is likely that we will be successful in the quest for the ToE.

Conclusion

There is no such thing as time at the most fundamental level. Our perception of everyday things misleads us to believe that something flows in one direction. We have the impression that things or events are stable, fixed objects located in a specific place and a particular time. Physicists are baffled by the fact that time always points to the future; our view is that time is one direction. It never reverses even though all the laws would work just as well if time ran backward. Some physicists argue that we should give up the concept of time to merge general relativity and QM.

We learn from QM (entanglement and nonlocality) and consciousness that there is no separation between things; we are all connected. There are no separate points in space. Time is simply the memory of previous events. Past, present, and future are all interlinked. Memories, cycles of lives, and so on, should not be confused with time. Our perception makes things appear and disappear. Present, past, and future all exist simultaneously. The future only manifests when we observe it or create it.

We call time the process of change; what goes on in physics is that a change occurs, such as moving from point A to point B. But the laws of physics are timeless, immaterial, and unchanging. It is clear that if there is no change taking place, we will not talk of time. Therefore, in an ordinary sense, time is equal to change. Time is something we have created to describe change taking place. It appears that the concept of time in physics is relative and depends on which field of physics one is referring to. But the overwhelming realization is that time describes the change.

The dominant paradigm in physics that explains everything as made of matter can no longer stand, as overwhelming evidence is against it. Many of the puzzles in physics can no longer be explained in terms of matter, space, or time alone. They cannot be described in terms of matter and energy only, in a nutshell, by the assumption that the basis of our reality is physical. Our reality is something else. Information and consciousness must be considered. Through some of its experiments, QM teaches us that we must give up the linear notion of present, past, and future and the notion of linear time.

A multidisciplinary approach to time should help us grasp the concept. For instance, during a dream, consciousness can navigate past, present, and future instantly. The distinction becomes nonexistent. This simply means that it is possible to navigate a timeless, multidimensional environment. For instance, I had a dream that I was invited to attend a Washington conference on the technological development of the Democratic Republic of the Congo. Although I have never been to the USA before, one night, six months before the conference, I could see myself in a timeless, multidimensional environment in Washington, talking and addressing the audience.

A few months later, I went there and gave a talk, and during my stay, I could recognize some of the places I visited, even where I stayed, because my consciousness had already seen these places or had access to them. This dream, like so many others, makes me argue that time is a tricky concept; it is not an exact notion and may not be an appropriate word to use. I hope that understanding time in a new frame will bring a breakthrough. Somehow, time appears as our construct to navigate this environment, in the same way that we tend to think of particles as physical objects. This picture is highly misleading. Table 3.1 shows various approaches to time and their meanings.

It is clear now that time has different meanings in classical mechanics, QM, QFT, and other branches. Wheeler and DeWitt, in their effort to unify relativity and QM, suggested that time does not exist and that, at the fundamental level, the universe is timeless. Many other theories, such as ST, explain that space is emergent, leading some researchers to argue that if space is emergent and approximate, time should be the same. It is clear now from the discussion in this book that space and time are not some precise physical location (area). It appears that if one understands the role that consciousness plays in the universe, which is a fundamental or primary role, the concepts of time and space become meaningless or disappear entirely. We start learning that we live in a timeless, multidimensional environment.

Table 3.1   *Approaches to time and their meanings*

Approach	Meaning of Time
Everyday Meaning	Time is about measurement during a period in which something happens. That is why time is divided into second, minute, hour, day, week, month, and year, solely to make sense of the world, to keep a record.
Prof. Carlo Rovelli	Time does not exist; it is an illusion. We should revise our understanding of it. There is no time variable in the fundamental equations that describe the world.
Dr. Julian Barbour	The concept of time moving in one direction is an illusion. We inhabit a universe with no past or future and live in an internal presentation. Time, as such, is an illusion and does not exist.
Classical Mechanics	The concept of an object or the path of a particle has a real meaning in classical mechanics. Regardless of whether something happens, whether change happens, time is passing; time passes uniformly.
Theory of Relativity	Time is not absolute and depends on the observer. Time is relative and dynamic; past, present, and future are not absolutes. Time in general relativity is not a universal constant.
The Block Universe Theory	The universe is static, and past, present, and future all exist simultaneously and are real. The concept of the arrow of time or time flow is simply a mental construct.

Approach	Meaning of Time
Quantum Mechanics	Time does not appear to be a fundamental element; nonetheless, it is there. For example, a proton does not experience time. Some recent work suggests that time does not exist at the quantum level, meaning that time is a function of relativity only.
	Also, the idea of nonlocality indicates that the linear passage of time does not exist; thus, time, as such, does not exist. The notion of time becomes meaningless.
	The Copenhagen interpretation of QM posits that the collapse of the wave function cannot be undone or reversed; the process is time irreversible.
Quantum Field Theory	An antiparticle has a negative charge. Antiparticles are ordinary particles, but they move backward through time. So, in QFT, the notion of particles moving back through time is a reality.
Consciousness	Waking consciousness: The perception of the flow of time results from our waking consciousness, in particular, how our brains process information to help us move around.
	Dreaming consciousness: In a dream, we can move in three dimensions of time (so-called past, present, and future) at the same moment. It is like space and time do not exist. We navigate freely, without any obstacle, in these temporal and spatial dimensions.

Although the idea of time seems controversial, I believe that we are likely to get a better view if we combine various subjects of knowledge, such as philosophy, psychology, biology, consciousness studies, and QM, to come up with a multidisciplinary perspective and approach to understanding time. Like other scientists, much of the work I do is work in progress, and future discoveries may change or guide the direction of, complement, or add to my research.

PART FOUR

The Role of Information In Understanding The Universe

What is Information?

We are beginning to see the entire universe as a holographically interlinked network of energy and information, organically whole and self-referential at all scales of its existence. We, and all things in the universe, are non-locally connected with each other and with all other things in ways that are unfettered by the hitherto known limitations of space and time.

—ERVIN LÁSZLÓ

I have used the term information throughout this book and mentioned it in the previous chapters. Nevertheless, I have not yet given a full explanation nor have I discussed it any detail. In this chapter, I would like to elaborate on the concept of information and explain its role in understanding reality in addition to its remarkable connection with consciousness. I would like to stress that information may be unique because it has started to occupy an important place in various fields of science and art. For instance, during the last twenty years or so, the concept of information has been used in multiple areas of knowledge, either as a metaphor or taking numerous other roles in explaining diverse phenomena.

In what follows, I define what information is and explain its laws. Also, I discuss the different types of information; the roles of information in multiple fields of knowledge, such as biology and physics, are emphasized. Moreover, I discuss an important and fundamental aspect of information: the creative process. I move on to show the special relationship that exists between information and consciousness. I finish this chapter with a summary, pointing out the fundamental role that consciousness plays concerning information.

I have used several concepts in this book, such as matter, energy, and time, to explain consciousness's importance. In this chapter, I add information. Why information? The answer is simply that several physics and biology researchers have started to recognize the need to use this concept to understand the universe. In the following, I will attempt to answer several questions. For instance, consider the following:

- What is the meaning of the concept of information in physics?
- What is information made of?
- What are the mechanics of information in physics?
- What role is information given in physics?
- What is its status? Is information fundamental or not?
- How can information contribute to our understanding of consciousness?

These are just some questions among many that I shall address in the following pages. Therefore, this chapter aims to present and explain the role of information in relation to matter, energy, time, and consciousness in understanding the universe. The key to answering these questions may be closely analyzing different types of information and their creative aspects. By looking carefully and understanding its creative part, one may see why this concept, along with consciousness, may be used in physics and may perhaps form the basis for answering several fundamental questions in the field of theoretical physics.

Furthermore, to answer these questions, an attempt to address how information is processed and computed at the atomic level is made. What kind of calculation goes on at this level? We must address, among other things, how matter registers information. Although progress has been made in the understanding of matter, energy, space, and time, very little has been made in our understanding of information. Understanding information can help not only physicists but also scientists in other fields to better understand the intrinsic nature of physics or other related subjects.

At the beginning of this book, I pointed out that our present understanding of atoms, electrons, matter, energy, and time is incomplete and needs a careful reexamination. It is only by reviewing and evaluating these concepts that we may gain, hopefully, new insights. It is possible that by analyzing and looking closely at these concepts, we may be likely to make progress in the quest to find the ToE. I firmly believe that information can provide us with new insights that may contribute to a new and better understanding of matter, energy, space, and time. The way I see it, information provides a functional explanation.

I am indebted to several researchers who have discussed different forms of information and have attempted to put forward information laws. These approaches to categorizing information and proposing its rules are a bonus to me, as I can dig deeper to understand its true nature. Without these developments, it would have been challenging to pinpoint its role in understanding the universe, in particular matter and energy. I have little doubt that information appears to be a powerful and essential tool in understanding several developments in physics, particularly theoretical physics. I would also like to point out that information has become an integrating factor, discussed in various fields, including physics, mathematics, philosophy, consciousness, biology, and psychology. Figure 4.1 illustrates the role of information as an integrating factor.

Information has various ramifications in almost all the fields of knowledge, ranging from anatomy to zoology. Also, the concept of information has become a unifying factor in science, where it can be used as a concept to explain several aspects of physics, biology, communication, and more. In a nutshell, it transcends frontiers. "Proposed by Boltzmann as entropy, information has evolved into a common paradigm in science, economy, and culture, superseding energy in this role. As an integrating factor of the natural sciences, at least, it lends itself as a guiding principle for innovative teaching that transcends the frontiers of the traditional disciplines and emphasizes general viewpoints" (Dittrich 2015, 1).

My intention is not to discuss the concept of information as used or developed in areas of science other than physics and biology. This has already been discussed elsewhere and is well documented, see, for instance, Kennedy (2014). A good portion of the discussion is reserved for the use of the concept in physics. I have discussed its use in biology because of what physicists can learn from living systems and the biological concept of information of living systems. The hope is that this will result in a better understanding of information and its use in physics. Hopefully, it may offer us valuable insights into the subject of this book and help us to better understand the two related concepts of matter and energy.

Also, I do not cover any theoretical or philosophical issues surrounding information theory and its implications. Several other researchers have discussed these already. For a useful review and a discussion of philosophical issues and their implications regarding the concept of information in biology and physics, see, for instance, Bates (2006) and Barbieri (2016).

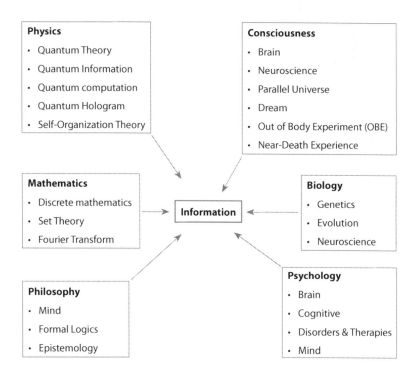

Figure 4.1 *Information as an integrating factor*

In showing the vital role of information in the universe, I demonstrate that matter, energy, information, and consciousness are all related. Information requires a carrier, such as a wave, book, or TV and a receiver/transmitter, such as a CD or DVD. The transmission of information needs a page to move it from one place to another. Some researchers have used information as a metaphor, see, for example, Gershenson (2012). However, some claim that information is the basis of reality, see, for instance, Lloyd (2006) and Wolfram (2002).

DEFINITIONS OF INFORMATION

We experience information in various forms—for instance, from sensory inputs and thoughts, or alternatively, from whatever we are in contact with. We interact with information through our five or six senses. For instance, if I look at a tree of tropical fruit, such as mango, I get information about the tree, such as its color, the tree's size, the form of the mango, and the smell of the fruit. There are many definitions of information, depending on the field one is discussing.

The word information" is not self-explanatory; information exists only as a potential and needs a medium to manifest. Hence, defining information constitutes a difficult task, but one thing we are all accustomed to is that information can be transmitted or communicated instantly. One can get direct information about a physical object's properties, such as a star in the sky or a tree located far away. Information is what an observer sees, hears, or touches. When we look at something, we receive different types of information. Therefore, the meaning of information depends on the subject or observer. The same word or message may have different meanings for different people.

A biologist will talk about biological information, which simply means a genetic program found in living organisms. On the one hand, a physicist may say that the universe is made of bits or qubits. A bit stands for a binary digit and is the smallest unit of data in a computer. A bit has a single binary value, either zero or one. A qubit stands for a quantum bit. In quantum computing, this is the basic unit of quantum information and is the quantum version of the classical binary bit that is physically realized with a two-state device. On the other hand, a chemist will talk about chemical information encoded by a structure, e.g., the atomic number of sodium is 11, and 11 protons and 11 neutrons surround it.

But other researchers have come up with perhaps a more fundamental definition of information. For instance, Bates and Parker define information "as the pattern of organization of matter and energy" (2006, 1033). Bates and Parker's definition seems to be the one adopted by many physicists, even though they do not refer to it directly. Some writers and researchers, such as Seth Lloyd, Tommaso Toffoli, Edward Fredkin, and John Wheeler, have developed and put forward the hypothesis that information is a fundamental ingredient of reality.

Why have some physicists taken this new direction? It is simply because the concept of information, considered either as a metaphor or as an ingredient of any physical theory, has answered several fundamental questions and allowed us to understand science in a novel way. There is no doubt that information and computation (the process of information processing) have a special status in physics. It appears that there are advantages to looking at information to understand the world rather than relying on the concepts of particles or energy. And this is true particularly in the study of physics, biology, psychology, and chemistry.

Laws of Information

If some physicists or researchers believe that information is a fundamental ingredient in physics, we must understand the laws that govern it. Because we talk about information, we might ask ourselves, are there any laws of information similar to other laws of physics? Over the years, attempts have been made to show through observations and experiments that these laws exist. These laws are about how information behaves, how it interacts with matter, energy, or other things. Below, I have summarized the laws that govern information, based on the work of Carlos Gershenson. These laws will help us understand if our universe is made of information and how best to use the concept of information.

This quest to understand information and its supposed role in running the universe has led several writers and researchers, such as Carlos Gershenson, to put forward eight tentative information laws. These are the laws of information transformation, information propagation, requisite property, information criticality, information organization, information self-organization, information potentiality, and information perception.

In what follows, I discuss these laws and their implications. Understanding these laws opens our eyes to the true nature of information.

LAW OF INFORMATION TRANSFORMATION

Gershenson says that information is not a conserved quantity that can be created, destroyed, or transformed. Information is altered during the interaction. In his words, "information will potentially be transformed by interacting with other information" (2012, 5). The key word here is "transformation." This is because when information is out there, it is performing computations. And during these calculations, various operations, such as addition, subtraction, and multiplication, are conducted, resulting in change.

LAW OF INFORMATION PROPAGATION

Gershenson also discusses the propagation of information. Information is spread by various means of communication. He says, "Information propagates as fast as possible" (2012, 6). He discusses the ability of different types of information to propagate. Among the most well-known are biological information, chemical information, and physical information. For example, biological information is passed between living entities in the form of DNA. In addition, information is shared instantly in the universe, carried by radio waves and sound waves, to name only two.

LAW OF REQUISITE COMPLEXITY

Based on the law of information transformation, Gershenson points out the ability of transformed information to increase, decrease, or simply remain constant. He argues that "More complex information will require more complex agents to perceive, act on, and propagate it. [Alternatively], the law of requisite complexity just states that the increase in complexity of information is determined by the ability of agents to perceive, act on, and propagate more complex information." (2012, 7).

LAW OF INFORMATION CRITICALITY

Gershenson writes, "Transforming and propagating information will tend to a critical balance between its stability and its variability" (2012, 7). He stresses a contrast between propagating information and transforming information.

The former is about a continuous transformation of information, while the latter is about being stable (stability).

LAW OF INFORMATION ORGANIZATION

"Information produces constraints that regulate information production" (Gershenson 2012, 8). It is about how information creates limitations or restrictions that control the output of information. Evolving information regulates information production.

LAW OF INFORMATION SELF-ORGANIZATION

Self-organization is the physics of information processing in complex systems. Chris Lucas (1997) defines self-organization as the evolution of a system into an organized form in the absence of external pressures. "Information tends to its preferred most probable state" (Gershenson 2012, 8). Self-organization is achieved in nature through different means. For instance, in natural phenomena, such as biological systems and physical systems, information organizes itself. Self-organization, as I understand it, shows the presence of intelligence, a lower form of consciousness that helps organize everything. Consciousness is at the center, though it is often not recognized. Information in itself cannot do anything. Apart from its movement, carried by energy, it needs intelligence to organize itself. Consciousness is behind this self-organization.

LAW OF INFORMATION POTENTIALITY

"An agent can give different potential meanings to information" (Gershenson 2012, 9). The same information can have different meanings depending on the receiver or agent. And the purpose of information depends on the context of observation.

LAW OF INFORMATION PERCEPTION

"The meaning of information is unique for an agent perceiving it in unique, always changing open contexts" (Gershenson 2012, 9). The author thus stresses the unique importance of information for each receiver.

Types of Information

Kennedy (2014a) differentiates two concepts of information: symbolic information and physical information. The former refers to the everyday use of the term information, which focuses on data, history, student's knowledge about their subjects, knowledge of foods, actuality, political knowledge, social understanding, economic news, and so on. Physical information, however, is mainly used in science—for instance, the concept of entropy in thermodynamics. Kennedy (2014b) stresses that it is important to distinguish between these two types of information in our quest to understand the meaning of information. He insists on making this distinction to avoid any confusion about two different concepts of information.

Information is found in matter, in energy, and throughout the universe. Prof. Marcia J. Bates, as mentioned above, discusses the fundamental forms of information. She writes about four types of information: natural, represented, encoded, and embodied. However, she insists that all information is natural information. For a full discussion, see Bates (2006).

Natural

NATURAL INFORMATION

Bates says that all information falls into the category of natural information and that this kind of information exists in matter and energy.

REPRESENTED INFORMATION

Represented information is merely natural information that is encoded or embodied, such as information found in biological or living systems—for instance, animals and plants. This kind of information can be found in association with living organisms.

ENCODED INFORMATION

This type of information includes symbolic, semantic, and rhetorical information and telecommunications.

EMBODIED INFORMATION

This refers to the actualization of information, which was previously in encoded form. This is exemplified by information found in the biological system, such as in plants and animals.

PHYSICAL AND BIOLOGICAL INFORMATION

Some authors, including myself, prefer to classify different types of information according to the field of study—for instance, biological information, chemical information, and physical information. In the following pages, I discuss biological information and physical information.

BIOLOGICAL INFORMATION

One of the constituents of all living beings is DNA. All information on the building blocks of living beings is encoded in our DNA (see Figure 4.2). Usually, it is referred to as biological information. The genetic code and DNA teach us about biological information and the processes that go on.

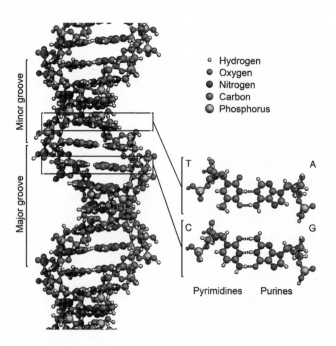

Figure 4.2   *The structure of DNA, showing in detail the design of the four bases—adenine, cytosine, guanine, and thymine—and the locations of the major and minor grooves. Image by Zephyris via https://commons.wikimedia.org/wiki/File:DNA_Structure%2BKey%2BLabelled. pn_NoBB.png (Creative Commons Attribution-Share Alike 3.0 Unported license)*

By looking closely at biology, we see that living systems process matter and energy and are imbued with DNA information, and DNA shows remarkable intelligence in processing information. When we talk about intelligence, for instance, cell intelligence, the idea of consciousness comes to mind. This intelligence is a form of consciousness. Thus, at the fundamental level, all biological systems show a kind of intelligence, which is simply the presence of consciousness, thus emphasizing the role of consciousness in the universe.

Information in human beings or matter is stored in the body, such as the brain or the biological systems. An essential form of information, particularly in biology, is the genetic code, which is the key to understanding several aspects of living systems. Physicists could learn from biologists how DNA stores and processes information and apply it to studying the atom's tiny bits. This biological aspect of information has a lot to offer in the quest to learn how, for instance, genetic information is transmitted.

One of the important points about biological information is the process whereby genes are carriers of genetic information. Although the chemical paradigm in biology is still popular: most biologists support it based on the belief that "life is chemistry," the information paradigm in biology is gaining ground. Information-based processes are hereditary, such as natural selection. In biology, information seems to be a fundamental component of living beings such as plants, animals, and human beings.

Although both paradigms exist, my intention is only to point out this kind of information in living entities. What is suggested is that living beings are not only chemical entities that can be fully understood in terms of matter and energy but that living entities are information processing entities. For in-depth discussions of life as chemistry and life as information, see Kay (2000) and Smith (2000).

PHYSICAL INFORMATION

Information theory was born in 1948 when Claude E. Shannon published his book *The Mathematical Theory of Communication*. He outlined his views of information, particularly the concepts of Shannon entropy. Shannon was inspired by the Boolean algebra invented in 1847 by George Boole. Boole's fundamental concepts are logical operations, such as *and, or,* and *not.* Over the years, these concepts have been used in many branches of science and technology. Shannon, Toffoli, and Margolus used it many years later to represent components in electronic circuits, such as switches and relays.

Shannon is regarded as the first scientist who introduced and defined the concept of information; he described the most basic information theory, which is simply the binary digit (bit). The unit of information is the bit. The bit is simply a choice between two alternatives. Binary refers to something that consists of two parts: zero/one, black/white, in/out, or hot/cold register as a bit. He showed that information could be measured and quantified.

Although the work of Shannon is well-known, I would like to stress that over 12,000 years ago, Africans developed the binary code, referring to it as the Ifa divination system; its significant similarity with computer science concepts is well documented (Lokanga 2020). Table 4.1 shows the system's punches and their corresponding Yoruba names. The Yoruba people are an ethnic group that inhabits Western Africa, mainly Nigeria and Benin. For a detailed discussion of the African binary system, see Alamu et al. (2013).

Table 4.1   *Shows the punches and their corresponding Yoruba names*

Sequence	Punches	Ifa (Odu) Name
1	000000	Ogbe
10	000001	Osa
13	000010	Otua
14	000100	Irete
9	001000	Ogunda
5	000011	Irosun
15	000101	Ose
3	001001	Iwori
4	000110	Odi
16	001010	Ofun

Sequence	Punches	Ifa (Odu) Name
6	001100	Owonrin
7	000111	Obara
11	001011	Ika
12	001110	Oturupon
8	001110	Okanran
2	001111	Oyeku

In information theory, any variable that can assume the values zero or one, yes or no, is called a binary digit or bit. We can use a binary number, such as zero or one, to transmit a set of instructions. Instructions to make a binary choice are simply given by sending, for instance, one to suggest hot or zero to suggest cold. Briefly, several bits of information can be encoded in a physical system; this is true when, for example, a set of instructions in the form of n binary choices need to be transmitted to identify the physical system's state. Information can be encoded in biological systems. Physicists have concluded that any information can be encoded, processed, and finally transmitted by physical means.

This development has led some physicists to posit that information is fundamental. They claim that the universe is made of information. Although many physicists have accepted information (physical information) as essential without looking deeply at its intrinsic nature, my understanding is that the concept of information offers us a novel way of thinking about physics and its components as well as the universe.

Physicists have realized that information is fundamental in physics and is a component of atoms and electrons. In contrast to chemists or biologists, who argue that information is present only in living systems, physicists consider information to be present in atoms and electrons. Electrons or subatomic particles process information, have memories, think, and behave like intelligent entities. Electrons behave like biological information systems, living entities. Although considered as non-living, atoms store information.

Everything in the universe contains information: atoms, electrons, trees, and plants are all carriers of information. Furthermore, information in physics refers to physical information, which is the physical system's information. For instance, the mass of an object, its size, its color, and its characteristics are physical information. The other use of the concept of information in physics comes from thermodynamics. Some physicists use the words information and entropy interchangeably. Entropy has, for many years, been described as a disorder, but more recently, physicists have realized that what we call disorder is actually information—information is equivalent to entropy. If one says that a system has more disorder or less disorder, this should be taken as meaning more information or less information.

The other meaning of information in physics has been influenced by the approach developed by the late physicist John Archibald Wheeler. He put forward the concept of the participatory universe, according to which physics can fundamentally be reduced to a yes or no binary choice: "Otherwise put, every 'it'—every particle, every field of force, even the space–time continuum itself—derives its function, its meaning, its very existence entirely—even if in some contexts indirectly—from the apparatus-elicited answers to yes-or-no questions, binary choices, bits" (Wheeler 1989, 310).

The work of Claude Elwood Shannon has been important to this development. Moreover, I firmly believe that the development of information technology and computing, coupled with advances in information theory and the technology of digitization, have all influenced physicists. Besides, in trying to find answers to the many unresolved conceptual issues in QM's foundations, many thinkers, particularly those in the physics community, have started to think of new ways of understanding all physics as a form of computation. The rapid growth in two fields of physics, namely quantum information and quantum computation, has heralded a novel way of understanding QM. And to some extent, these support the view that information is fundamental and should play a crucial role in our understanding of our universe.

One form of physical information is quantum information. This is used to describe quantum phenomena, such as quantum entanglement, nonlocality, and quantum tunnelling. In physics, quantum information is the key to understanding the concept of information. Quantum information has different properties from classical information. The bits that compose classical information can be either zero or one. On the other hand, the qubits that makeup quantum information exist in the superposition of the zero and one states. This shows that the information of a quantum system is different from the information content of classical systems.

By now, from the quantum computer or quantum information, we have learned that atoms, electrons, trees, and everything in the universe compute, meaning that they register, process, transform, and emit information. According to Prof. Seth Lloyd, "The laws of physics determine the amount of information that a physical system can register (number of bits) and the number of elementary logic operations that a system can perform (number of ops). The universe is a physical system" (2001, 1).

Although Lloyd may be right, my reading is that, as shown throughout the book, many physicists have not been vocal about the role of intelligence or consciousness in particles, minerals, plants, and animals. The universe is an intelligent, self-computing consciousness—a smart information processing system. There is no doubt that particles, for instance, in a double-slit experiment process information and have information processing capabilities comparable to those of an intelligent living entity.

The same is true of atoms and molecules, which also show a degree of intelligence. There are relationships between matter, energy, information, and consciousness. Without consciousness or intelligent life, an entity, such as a particle, will not interact or move around. It is like a coordinated, well-planned movement, obeying some laws.

I would like to observe parenthetically that information is found everywhere—in physical objects (e.g., the moon) and in biological systems (e.g., DNA and, cells). However, we have other types of information that do not need matter or energy as a medium of communication, for example, information received in a dream, which is nonlocal information. Yes, indeed, not all information is physical. This realization leads me to look closely at the definition of information.

M. J. Bates and Edwin Parker define information "as the pattern of organization of matter and energy" (Parker, 1974, 10; Bates, 2010). Extending this definition and applying it in physics to understand atoms, energy, and matter opens our eyes and helps us to understand the role of information in physics, which starts to become apparent. In natural or living systems, information behaves like consciousness. Should we not be saying that consciousness is the pattern of organization of matter and energy? I would like to think so. It is consciousness that is the source of everything; it is consciousness that appears to be the pattern of organization of atoms, electrons, and everything that exists.

The Creative Process of Information, and the Role of Consciousness

Creativity is a fundamental aspect of computation or information processing in living entities like plants, animals, or human beings and in QM. When atoms collide, they exchange information, resulting in a new arrangement and various other outcomes depending on the elements involved. Likewise, the exchange of information between particles generates new particles and creates and displays different behaviors. Information processing in matter and energy is creative: creativity is at the center of information processing.

The computation performed between atoms—or between atoms and electrons—effortlessly creates new arrangements, new things, and new configurations. New systems come out of the complex computations taking place. This is because atoms have embedded programs, i.e., specific instructions to behave in a certain way. The outcome of any reaction is fully known in advance. It is like each atom was programmed at the universe's inception with a well-defined mission. But it is important to stress that it is consciousness that tells information through computation what kind of pattern to create. Once again, this invisible force (which is simply consciousness) moves everything in an orderly way to create new forms and designs.

Electrons in an atom are embedded with information and continually work together in an intelligible way. It is possible that each electron or group of electrons has its role. The ongoing communication among electrons is fundamental to communication through the release of information into the environment or around the universe. The universe detects the released information, leading to the creation of a substance through an operation, such as addition, subtraction, or multiplication. The information is then released into the universe. The process continues and is ongoing throughout the day, eternally.

Trillions, billions, or perhaps millions of atoms are releasing information. Electrons do not only release information, but the process occurs in both directions: emission and absorption. They fire and absorb information continuously. This ongoing communication between atoms or particles (fields) is merely creative information processing. It is possible that without the information imbued in electrons or atoms, we would not witness the creation of various substances, and life would not be what it is. Communication between atoms and electrons (fields) is the key to the running of the universe. The nature of those entities is to disseminate, share, absorb, emit, and process. The information processing is the key.

During computation, waves carry information to different places and environments. Physical systems register this information. Alternatively, they go through the process of absorption and emission. The process is ongoing, meaning that other atoms or biological systems receive information, ready for another absorption/emission process; the ongoing process is continually taking place, as is the information exchange between the particles and the universe. Through this description, there is no doubt that information is becoming a fundamental component of physics, in addition to matter, space, and time.

Information is stored in matter, and energy and is transmitted throughout the universe. Different media are used for sharing information, such as soundwaves, DNA, human beings, TV, and radio. Information in matter is encoded and transmitted through waves; it guides and creates patterns. Information appears to be a blueprint that makes other things. Although information seems to be fundamental, it is consciousness that is the key, the master program that guides and controls the distribution of information, energy, matter, and particles in the universe. The running of the universe is dependent on the presence of consciousness.

I have realized that the most crucial property of information is creativity; this is the ability to create or the potential to reproduce. Information on its own cannot do anything. Consciousness is the creative force that merely consciousness confers information (natural information) with its creative ability. By elucidating the mechanisms of information transmission and computation and the role of consciousness, it is possible to build a trustworthy and reliable foundation for better understanding matter, energy, atoms, and subatomic particles.

Matter, energy, and electrons are not enough in themselves to provide a complete explanation of the physical world. Through an understanding of how the physical system uses information, the role of information and its creative ability becomes clear in physics. The transfer of information by a physical object or particle helps us in various ways to understand what is beyond matter. It is clear now that information is one of the fundamental physical entities. Physics must include information as one of its essential objects.

Information generates diversity, evolution, and adaptability; it creates new particles and new configurations under the guidance of consciousness. Information manipulates matter and energy to create new particles and new forms of matter. Information and computation (information processing) are independent of space–time and are found everywhere. The majority of scientists do not yet grasp the full implications of the role of information. Through information and consciousness, the universe is continually processing and creating. The creative process of information is a reality. The creative process, through an exchange, is ongoing.

Although the critical role of information has been highlighted, matter, energy, and information are also part of consciousness. Consciousness is the source of everything, the leading force. James E. Kennedy stresses this process of creativity or propriety of imagination shown by information: "Creativity is an important aspect of information processing in living systems that is often overlooked in discussions about the basic nature of the information. Creativity involves the generation of new conditions and behaviors, and is a fundamental property of life on all levels—from the evolutionary adaptations of individual living cells through human imagination producing technology and art" (2014, 1).

It is well documented in physics, particularly QM, how virtual particles continually pop in and out of existence. To my understanding, this are simply due to the fact of the creative process of information and computation. Throughout the book, I have insisted on the creative processes of communication and analysis. It is now clear that everything in the universe—atoms, molecules, particles, and living and non-living systems—is imbued with information. In fact, it is consciousness through computation (information processing) that organizes everything in the universe. The whole universe is based on information processing. This section helps us to understand the creative properties of information imbued in matter and energy. In fact, in everything, recognizing these unique properties allows us to understand the vital role of information in physics.

The nature of information found in the universe's physical system appears to be an essential starting point for understanding and decoding the nature of information itself. Many outcomes manifest themselves after computation. The role of both information and consciousness is emphasized in this book. For instance, regarding the measurement problem, well documented in physics, we can see that information plays an active role in every experiment and measurement. The reason is simply that those QM entities or subatomic particles are imbued with information; the computation is continuously taking place, whether or not the experiment is ongoing. This computation is directed by consciousness.

The interconnectedness of everything in physics appears to be due to the shared information or field of consciousness. Particles such as electrons, have a lower form of intelligence embedded in them. An intelligent system embedded with information responds to consciousness. Particles or fields have memories and the ability to process information. Let me say it once again: particles process and store information. The information processing capabilities of these entities is a reality.

This new understanding of the concept of information has made it possible to either use it as a metaphor or give it a fundamental role as a component of an atom. Many concepts in classical physics and QM, such as matter and energy, can be given a new meaning or interpretation in terms of information. The area of QM has been at the forefront of this development. Here, the concept of information has been used to understand, for instance, entanglement, nonlocality, quantum information, and quantum computing.

Summary

It appears likely that the unifying concept of information offers us a tremendous opportunity for solving some of the most pressing issues facing science and, perhaps, humanity. Looking closely at several issues facing theoretical physics, one realizes that it is an interdisciplinary approach that is likely to offer the necessary skills to researchers to be able to unlock the missing puzzle in the quest for the ToE.

Information and computation (information processing) are independent of space–time and are found everywhere. The majority of scientists do not yet grasp the full implications of the role of information. This chapter has discussed the concept of information as well as other related ideas, such as computation. Hopefully, this discussion has given us insights into the role of information in physics and the universe. Physicists need to rethink their approach to matter and energy. The current view is based on an incomplete paradigm. The new method compels us to consider the role of information and consciousness. One of the most critical elements of information is its creative aspect, through the ever-ongoing computation.

How is information transfer achieved? What is the role of consciousness in running the universe, the atoms, the fields? How is coded information in atoms transferred? Of course, during the collision, information exchange is performed, and various operations are conducted. But I must admit that there are still many things that we do not yet understand completely, such as the physical mechanism of information transfer between particles, systems, or objects. Some of the work has been done by Prof. Seth Lloyd, Dr. Stephen Wolfram, Prof. Tommaso Toffoli, Prof. Edward Fredkin, Charles Bennet, Rolf

Landauer, and others, but they have all neglected the role of consciousness. Hopefully, future work will show how this is achieved.

The theory of information discussed in this chapter suggests a wholly new direction for physics, a new way of looking at the world. Perhaps physics should learn from biology—for instance, with respect to how information is transferred, processed, and coded in the universe. Matter behaves like hardware and has some form of consciousness. Physicists need to develop a new approach, a new attitude towards the concepts of information and consciousness and their use in physics. I have addressed the issue of information in biology and physics. The purpose of information has been revealed. This provides us with the foundation for a theory of consciousness. I have stressed that consciousness is everywhere, which leads me to discuss the universal field of consciousness.

PART FIVE

The Universal Field of Consciousness

Consciousness as a Field

Think of consciousness like spacetime—a fundamental field that's everywhere.

—ANNAKA HARRIS

In this chapter, I have no intention of attempting to resolve all the issues raised in most, if not all, consciousness theories. My only aim is to examine the importance of consciousness and its role in the universe and particularly in physics. This is because the arguments developed in the last four chapters push us to look at consciousness's part. Evidence has been presented throughout the book showing why I think that consciousness is fundamental. In a nutshell, a review of various papers leads to the postulation that consciousness is a quality or attribute of quantum-like processes. Moreover, new theories of consciousness posit the existence of the so-called unified field of consciousness. This realization leads us to ask important questions regarding its similarities with the concept of a unified field theory in physics.

Various consciousness theories have been well documented, ranging from psychology, philosophy, neuroscience, and electromagnetism to quantum physics. For instance, recently, there has been an increase in approaches using quantum mechanics and electromagnetism to explain consciousness. This new landscape is producing various viewpoints. For example, some of the theories argue that consciousness is physical. In contrast, others stress that consciousness is immaterial. Neuroscientists view consciousness as a natural state of the brain.

I would like to start with a broad definition of consciousness, followed by a discussion of some general introductory concepts of the unified quantum field theories: panpsychism and individual consciousness. I will touch on the idea of interconnectedness and discuss how to resolve the combination problem. After this short introduction, I will discuss the two main groups of field theories: the electromagnetic field theories of consciousness and the quantum unified field theories of consciousness. By choosing these two models of consciousness, I would like to show that physics could be understood in a new way through the role that consciousness or consciousness fields play in the universe's running. By doing this, I also present the fundamental principles to facilitate the understanding of these new approaches.

In this quest, as in many others, several questions arise. The aim is to answer these questions. For instance, consider the following:

- Is there an individual consciousness and a separate collective consciousness?
- Does an underlying field connect us all?
- Is matter simply imbued with consciousness, or is it part of this universal field of consciousness?
- What is the unified field of consciousness?

First of all, let us look at some definitions of consciousness.

Jeanne Ball (2017) defines collective consciousness as "the overall social atmosphere that arises from the thought and behavior of all the individual members of a community or society." This line of reasoning and research is supported by the work of Di Biase, who stresses that "Information and consciousness are an intrinsic, irreducible and non-local property of the universe, capable of generating order, self-organization and complexity" (2019, 81). On the other hand, Stonier defines information as "the cosmic organizational principle with a 'status' equal to matter and energy" (1997).

The work of these three researchers compels us to think about the fundamental role of consciousness. They say that there is an intrinsic relationship between matter, energy, information, and consciousness. Also, they explain how consciousness behaves in relation to information, matter, and energy. In fact, consciousness is the primary force, the organizing entity that runs and organizes everything in the universe. In this way, one can understand the universe through consciousness. It is a self-organizing, self-creating, self-maintaining entity or system.

The last few years have seen some researchers thinking of consciousness as a field that extends throughout space and the entire universe. For instance, in 2006, Bernard Haisch suggested that quantum fields that permeate all empty space may be responsible for producing and transmitting consciousness, which emerges in complex systems. Consciousness thus becomes or can be understood as an intelligent system or entity. In this respect, consciousness is the source of everything, organizing information, guiding energy, and directing all forms of interaction in the universe.

Now, if consciousness is everywhere, does that mean it is a field? I would like to say yes. I talked about various fields that carry information, energy, and matter. These fields behave like an intelligent, conscious system. In other words, they are consciousness itself. It is everywhere. This idea that consciousness is all around us is similar to the concept of a hologram. In a nutshell, the main idea of a hologram is that all the system's information is in every part of the system. Even if you cut an apple in half, each half contains all the information found in the other half. We can better understand this work through Karl Pribham and other researchers working on the holographic universe theory.

Karl Pribham's work on the brain field suggests that thought, image, color, and memory are created through quantum mechanical processes and that memory is not stored in our brains, but thoughts, ideas, smells, and images are stored outside the brain, and memory may be located in a quantum field or wave field, leading to the hypothesis that consciousness itself is part of this quantum field.

In my discussion about QFT, I mentioned that a particle is considered a field. That is why we talk about the electron field, quark field, or other types of fields. These fields carry all kinds of information and behave in a smart, intelligent way, as though they were conscious entities or intelligent systems. This field, alternatively called ZPE, implies a web of interconnectedness among all the universe's entities. When we look closely at the ZPE concept discussed throughout this book, it becomes clear that we receive information from this quantum field, which communicates with us continuously.

Several other writers have discussed the location of the individual or collective consciousness. Where is this consciousness located? According to J.J Hurtak and Desiree Hurtak, "Consciousness is suggested to be directly created by individual interaction with the non-local, all-prevalent quantum field. Our retrieval of information establishes within us a second-order quantum field or 'Mind-2,' which can acquire higher insights and knowledge through this function" (2011, 24).

The importance of understanding this field or quantum field is a challenge that we are all trying to come to terms with. One of the keys, pointed out earlier, is the concept of absorption and emission. This proves that emission and absorption seem to constitute a network or networking field where we are constantly exchanging information and, at the same time, processing it. Our body works like a biological computer (biocomputer) or quantum computer. Information processing is an ongoing process; thus, we can argue that we live in this considerable quantum field, where information is continuously being exchanged. We are thus connected to this quantum field and are able to tap into or join this vast sea of information at any time; we can relate to this spaceless and timeless environment.

One of the ways in which we can understand this interconnectedness is through dreams and remote viewing. During dreams, our dreaming consciousness has access to and can navigate an interconnected, timeless universe. For instance, through a dream, our consciousness can view and amass information without any effort; it can access a timeless universe. It has instantaneous access to an infinite amount of information, regardless of where it is located. In a way, we can argue that our consciousness is connected to this quantum field.

Commenting on the issues of dreaming and remote viewing, J. J. Hurtak and Desiree Hurtak argue that "These events, or 'occasions of experience,' could be considered actual quantum state reductions of events in physical reality that were accessed through the quantum field. This further suggests that the quantum field is also a consciousness field that does involve quantum state reductions, e.g. in a form of quantum computation whereby the larger mind functions like an Encyclopedia Galactica in the universe" (2011, 25). It follows that dream and remote viewing help us establish a theoretical relationship between the quantum field and consciousness.

Matter carries information through the wave; it is connected to this vast quantum field. This field contains everything—information about you, me, plants, animals, and non-living things. At the same time, this field is a force, energy, and information field. This demonstrates the intrinsic relationship between matter, energy, information, and consciousness. There is no doubt that this is one of the plausible explanations of how everything interacts in the universe and particles in the form of energies interact with other things in the universe to effortlessly create new arrangements or items.

It appears that the existence of the consciousness field, quantum field, or universal field of consciousness is something we ought to take seriously. It seems that the mind as part of a quantum consciousness field is primary; it seems that everything in the universe is connected. No wonder consciousness exists in this field, meaning throughout the field. This consciousness is embedded in the universe and in all entities, manifesting itself sometimes as an expression of force, energy, and information.

One of the recurring questions has been about the location of consciousness. In the consciousness field or quantum field model, consciousness is not located in the brain. It is rather a field that is received by living and non-living entities. Consciousness is received through various antennas situated in our bodies. One of them is the brain. What about individual consciousness? Although this individual consciousness exists, each of with this embedded consciousness belongs to the universal consciousness, similar to a drop of water (taken as an individual consciousness) merging with a river or an ocean (universal field of consciousness).

This realization has led many researchers to reflect on the true nature of reality. For instance, Jeanne Ball asks, "If this non-material field is the essence of everything, is it also the essence of individual and collective consciousness? If the unified field is at the core of everything – even at the core of our bodies – is it also hidden deep within the mind?" Yes, it is imperative that we create and promote harmony, that we spread love, unity, and peace among all beings because we are all connected through an underlying field. Everything any one of us does affects the rest of humanity.

There is a direct link between consciousness and the unified field. There is an underlying field unifying both consciousness (mind) and the physical (matter). More and more people have recognized that consciousness does not arise in the brain but is a separate entity in itself, an entire field. Understanding that we are all interconnected on a deeper level reminds us that we have to come together and work to benefit humanity. We must make a meaningful contribution to humanity, building a better community and working for a harmonious society.

We can now understand energy and information in terms of fields. It is possible to talk about the energy field and the information field. So, these fields play, without a doubt, an essential and fundamental role in understanding physics; they also indicate the role of consciousness. The existence of these fields adds a new dimension to physics and challenges other approaches to physics. We are all connected to this unified field of consciousness. This field is an intelligent system: it behaves like a smart system. Also, one of its properties is creation. This field is the basis of all design.

In the universe, we have various fields: electromagnetic fields, quantum fields, biological fields, ZPE fields, and more. For instance, the existence of biological fields is well documented in the work of Attila Grandpierre (1997), who pointed out their presence in cells. They are also known under the name of morphogenetic radiation and are related to the electromagnetic field. All these fields interact regularly and bring us different outcomes. These fields interact with individual consciousness as well as the collective or universal consciousness. For instance, it is possible that when electrons absorb and particularly when they emit, the information, vibration, and energy spread through the field of consciousness, perhaps similar to how sunlight radiates throughout the entire universe. This behavior of consciousness shows its quantum nature.

The information about everything that exists in the universe is located or exists in this quantum field. All actions are situated in this immense universal or collective field of consciousness. So, for a long time, we thought that we knew what matter was, but the truth is that we need to review our understanding of matter before even looking at consciousness. If we do not fully understand matter, how can we know the nonmaterial? A starting point to unlock the secret of matter is to assume that it is a nonmaterial entity, as discussed throughout this book.

Those who study biopsychology, physics, and consciousness have applied quantum concepts, such as entanglement, complementarity, and the dispersive states and know that these play some role in mental processes. The overall feeling is that using quantum mechanics to understand consciousness may open the door to a better understanding, as much of the research conducted in this direction appears to be promising in the quest to decode these complicated issues.

It is clear now that consciousness does not arise from the brain nor does it arise from the brain processing information. Once more, it does not arise from matter. The answer is that consciousness is an intrinsic property of matter or fields. Everything in the universe is imbued with consciousness—plants, animals, stones, and electrons. Consciousness is embedded in these entities and is a fundamental property of the universe. In many of these entities, consciousness is there in a latent form.

Although many panpsychists adhere to this view, several questions arise and must be answered if we are to understand what panpsychism is. One researcher who has looked closely at questions regarding panpsychism is Annaka Harris (2020):

If the most basic constituents of matter do indeed have some level of conscious experience, how is it that when they form a more complex system—such as a brain—those small points of consciousness combine to create a new conscious subject?

For instance, if the individual atoms and cells in my brain are conscious, how do those separate spheres of consciousness merge to form the consciousness "I'm" experiencing?

What's more, do all of the smaller, individual points of consciousness cease to exist after giving birth to an entirely new point of view?

These questions are profound, calling us to reflect on this theory's implications, and they need to be answered if those who are not yet converted to the view of panpsychism are to be persuaded to look closely at this hypothesis. Many of the leading researchers have indeed had reservations about endorsing panpsychism. The solution for me may come from adopting a new view of consciousness, particularly from the quantum field prospects. The concept of fields is the key to unlocking consciousness and offering a better understanding of the idea. Although individual consciousness exists, the shared consciousness or field may be the key to embracing panpsychism and saying goodbye to the "combination problem."

As Annaka Harris rightly point out, the combination problem is "the hardest problem facing panpsychism."

The solution to the combination problem is to understand consciousness as a vast field that permeates the entire universe, analogous to the ZPE. In this way, consciousness is understood as a field that extends everywhere and interacts with both living and non-living things. This consciousness field is continually interacting with other fields. It is an intelligent, creative force, the source of everything, the most fundamental, the alpha and omega.

It is clear now that consciousness pervades the entire universe. Animals, plants, and minerals are imbued with consciousness, and they exhibit what I call an elementary or basic forms of consciousness. Matter, particles, and energy are brought together as a field that carries information and interacts with the environment. As rightly stressed by the physicist Freeman Dyson, "the processes of human consciousness differ only in degree but not in kind from the processes of choice between quantum states which we call 'chance' when made by electrons" (quoted in Hunt, 2020).

We encode information (sounds, images, voices, books, etc.) in electromagnetic fields; the information is transmitted to receivers, such as radio, TV, and computers. We receive information that reaches our brain from the outside because our nervous system or brain functions as a receiver and an emitter. This is how our body, brain, or nervous system tunes and receives consciousness. Our brain is a receiver and an emitter of consciousness.

There is this all-encompassing field of consciousness that exists. This unified, universal field of consciousness is the cause or the force behind all physical manifestations of everything in the universe. Each individual human consciousness is part of the field. It is possible to achieve the union of an individual consciousness with the so-called universal field, similar to how one drop of water merges with the ocean. The merging of an individual consciousness with the universal field is possible. Multiple consciousnesses exist and so does one universal consciousness. It appears that individual consciousness (mind) is part of this universal field of consciousness.

Electromagnetic Fields Theories of Consciousness

The electromagnetic theories of consciousness posit that consciousness can be understood as an electromagnetic phenomenon. Electromagnetic field theories propose that consciousness arises when a brain produces an EM field with specific properties. One of its ardent proponents, Prof. Susan Pockett, argues that "conscious experiences are identical with certain spatial EM patterns generated by neural activity in the brains of conscious subjects" (2012, 2). Different versions of the same theory have been put forward by various scholars over the years. But in this book, I only intend to discuss Prof. Johnjoe McFadden's conscious EM information (cemi) field theory and Tam Hunt's resonance theory of consciousness.

Some of the characteristics of the unified field of consciousness are the following:

- Consciousness is formulated as being identical to a field, similar to the use of the term "field" in physics.

- Consciousness has a duration and extension in space.

- Consciousness is suggested to be identical to some aspect of a physical field.

- Consciousness is a field.

Johnjoe Mcfadden:
Conscious Electromagnetic Information Field Theory

I came across some interesting work and a new approach using the electro-magnetic field concept to explain consciousness. This line of reasoning is presented by the work of Prof. Johnjoe McFadden, who teaches molecular genetics at the University of Surrey in England. He has come up with the cemi field theory of consciousness. McFadden tells us that consciousness is simply a complex electromagnetic field.

The proponents of the field theories of consciousness argue that conscious-ness is a manifestation of an underlying field or fields. Much of McFadden's work or theory is summarized in his book *Quantum Evolution*, published in 2001. He also wrote a book with Professor Jim Al-Khalili called *Life on the Edge: The Coming of Age of Quantum Biology*, published in 2015. As rightly pointed out by Tam Hunt (2019) in "Is consciousness just a complex elec-tromagnetic field?", McFadden proposes that "electromagnetic fields in the brain integrate our thoughts to generate our conscious mind."

McFadden suggests that the brain's electric and magnetic fields are possibly the seat of consciousness. In the same article (an online interview with Tam Hunt), he is asked whether "only EM fields are associated with conscious-ness or other fields as well" and replies that any dynamic field that encodes information could develop awareness and potentially become conscious.

One of the critical elements of this theory is the idea of the brain as a gen-erator of the electromagnetic field. It is well documented that the brain generates an electromagnetic field; this realization led to many applica-tions, such as brain scanning. Several researchers have built a theory of consciousness around this idea, including Popper, Lindahl, and Libet. As rightly pointed out by McFadden, Karl Popper held the view that "consciousness was a manifestation of some kind of overarching *force field* in the brain that could integrate the diverse information held in distributed neurones" (2002, 24).

McFadden proposes that consciousness is a manifestation of a force field in the brain. In a nutshell, he extended the work of earlier researchers and has perhaps gone as far as anybody in positing that the brain's electromagnetic field is the physical "seat of consciousness." In "Synchronous Firing and Its Influence on the Brain's Electromagnetic Field," McFadden explains the origin of the brain's electromagnetic field as well as its magnitude. He also reviews several works that led him to propose that the brain's electromagnetic field is the physical "seat of consciousness." The result is summarized by looking closely at evidence about the proposal and predictions he put forward.

Looking at the experimental evidence, we see that he has looked closely at several usages of electroencephalography (EEG), which is usually used to measure electrical activity in the brain. He points out that one of EEG's most critical features is that "the brain generates a highly structured and dynamic extracellular field" (2002, 4). For McFadden, the brain plays two roles at the same time. It is both a transmitter and a receiver of its EM signal. On the other hand, a thorough analysis led him to argue that endogenous electrical fields may influence the brain in several ways. The theoretical consideration is very sound; his arguments are well researched and backed by several highly sophisticated works and an analysis of the brain itself.

The cemi is based on three propositions:

- The brain generates its own field.
- The brain's EM field is the seat of consciousness.
- The brain's EM field can itself influence neuronal firing.

The cemi field theory's overwhelming conclusion is that it has come up with a possible solution to the binding problem. Moreover, it provides a new way of understanding consciousness. Several people have criticized McFadden and Susan Pocket, another proponent of the EM field theory of consciousness. Even though I disagree on specific points with McFadden, particularly on how consciousness arises, nonetheless, there are things I have learned from the approach he has used.

The idea behind McFadden's theory is that the brains of human beings and animals generate EM fields; thus, all brains are conscious. But he stresses that the EM fields present in most animals' minds appear not to have much of an influence on their brain activity. Thus, animals have a lower form of consciousness.

Other characteristics of McFadden's cemi field theory are the following:

- Thoughts are EM fields.
- External EM fields interfere with our thoughts.
- Consciousness is complex information in the bran's EM field.
- It predicts that complementary EM fields ought to be able to neutralize a thought.
- It suggests that conventional computers will never be conscious because the EM fields they produce are not part of their computations.
- In the future, computers that generate EM fields that contribute to their computation will be conscious.
- The theory allows nonconscious systems to possess awareness.

Tam Hunt: Resonance Theory of Consciousness

Tam Hunt and his collaborator, Jonathan Schooler, have developed a reso-nance theory of consciousness that posits that resonance or synchronized vibrations (see Figure 5.1) are the key to understanding both reality and consciousness. In several of their works, they use the terms "resonance," "synchronized vibrations," "harmonization," and "synchronization" to mean the same thing. Their approach describes everything in terms of vibration—matter, bodies, atoms, electrons, etc. Using well-known and well-documented basic notions of quantum physics, they point out that everything in the universe is continuously moving, vibrating, oscillating, and resonating at specific frequencies. Resonance is a type of movement characterized by swinging from side to side or between two states.

Hunt and Schooler rightly point out that everything in the universe comes down to a vibration of the underlying fields. They have used some of the concepts of the QFT theory, particularly the notion of fields, combined with the ideas of emission and absorption well developed in quantum theory. Consequently, they expand the phenomena of synchronization. When dif-ferent vibrations from matter or other things combine, they vibrate together at the same frequency. This phenomenon is well explained in the concept of holography. They stress that deep insights are gained from close analysis of synchronization. And this analysis leads to a deep understanding of the nature of reality and the universe as well as consciousness.

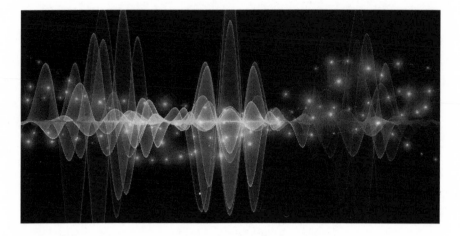

Figure 5.1 *An illustration of synchronized vibrations or resonance*

The resonance theory of consciousness was inspired by the work of the German neurophysiologist Pascal Fries (2022), who posits that consciousness results from the combination of various electrical patterns, such as gamma, theta, and beta waves, working in collaboration in the brain and producing different types of human consciousness. Therefore, consciousness arises in the brain from the interactions of various electrical patterns. Tam Hunt argues that the electromagnetic field is a possible seat of consciousness. He says that there are many other physical fields, such as gravity or nuclear forces, which may be the seat of consciousness in some systems.

I have noticed that in some respects, the work of Hunt and Schooler resembles the theory developed by Prof. McFadden. But the core of their arguments is based upon the work of Fries and several other researchers who did similar work before them. They accepted and promoted the idea that everything in the universe is imbued with some form of consciousness—a position known as panpsychism. The idea is that consciousness does not emerge from somewhere else but is always there, associated with matter.

Hunt and Schooler have made it clear that communication between different entities in vibrations is the key to expanding consciousness from a lower level to a higher level. This simply means that the expansion of consciousness depends on the type of communication between resonating structures. They continue as follows:

Accordingly, the *type* of communication between resonating structures is key for consciousness to expand beyond the rudimentary type of consciousness that we expect to occur in more basic physical structures. The central thesis of our approach is this: the particular linkages that allow for macro-consciousness to occur result from a shared resonance among many micro-conscious constituents. The speed of the resonant waves that are present is the limiting factor that determines the size of each conscious entity (2018b).

Hunt and Schooler's theory is all about vibration and oscillation. Resonance is at the center of human and animal consciousness and above all of physical reality itself. According to their theory, shared resonance is the answer to the nature of consciousness.

Quantum Unified Field Theories
of Consciousness

Despite the progress that has been made in unlocking the nature of consciousness, various questions remain. These questions have led some researchers to use quantum theory to decode consciousness. The link between quantum and consciousness was pointed out by the early pioneers of QM, and some experiments—for instance, the aforementioned double-slit experiment—have indicated the role of consciousness.

Some physicists suspect that consciousness has a vital role to play in understanding quantum phenomena. Some have even gone so far as to suggest that QM has a critical role in understanding the brain's workings. Some researchers have also discussed quantum effects in living beings, such as in human beings, brains, and plants. Some features of QM are regularly used in discussions about consciousness in the scientific literature, such as entanglement, complementarity and quantum randomness processes (spontaneous emission of light or radioactive decay).

There is no doubt that this use of quantum theory has led to some progress toward understanding consciousness. Quantum theory approaches have moreover been applied to explain the working of consciousness through different methods. Atmanspacher (2020) documents the three main procedures used.

In the first, consciousness is regarded as "a manifestation of quantum processes in the brain."

Second, "quantum concepts are used to understand consciousness without referring to brain activity."

Third, "matter and consciousness are regarded as dual aspects of one under-lying reality."

For an in-depth review of these approaches, see Atmanspacher (2020). In the following pages, I discuss three related theories of consciousness: John Hagelin's unified field theory, Stuart Hameroff and Roger Penrose's quantum consciousness theory, and Pim van Lommel's unified field of consciousness theory.

John Hagelin's Unified Field Theory

Prof. John Hagelin tells us that consciousness is a field or the unified field. He has demonstrated in an experiment that people meditating together have a positive effect on the rest of society, reducing the number of crimes in a particular city or location. In this well-documented experiment, the effects of consciousness were observed, analyzed, recorded, and most importantly, published in several journals (Tmhome 2016).

In the online journal Tmhome, Dr. John Hagelin argues for the importance of the unified field theories or super theories by stressing that they reveal the fundamental unity of life. He goes on to add that "They show that at the basis of our universe is a single universal field of intelligence, and the source of all order displayed throughout the universe" (Tmhome 2019).

Dr. Hagelin has explored consciousness through transcendental meditation and has realized that many theoretical models used in physics parallel another description of reality, one proposed by India's sages, based on exploring the structure of mind and consciousness. This more profound quest led him to realize that the description of the interior of the universe found in transcendental meditation "correspond[s] with remarkable accuracy to the discoveries of physics unfolding in my work" (Hagelin 2016). He points out the similarities between the mind and the world of quantum physics. For instance, the hierarchical structure found in the mind is similar to the one in quantum physics found in the study of atoms.

The similarities between the unified field theories of physics, based on the superstring, have been related to advances in neuroscience that have reported a unified field of consciousness. These two discoveries raise questions about whether there is an intrinsic relationship between the two or whether they are identical. John Hagelin has presented evidence that they are identical. For instance, Dr. Hagelin cites the aforementioned sociological experiment that took place in Washington. In "Is Consciousness the Unified Field?" (1987) he refers to a national demonstration project in which 4,000 people through meditation were able to reduce crime in Washington. This shows how we share a common field of consciousness, as those meditating were able to influence other people in the city. This suggests that human thought, belief, and awareness affect others through the common field. We all experience the unified field, whether we like it or not and whether we are aware or not.

Evidence supporting his work is found in a number of experiments showing what is called "the Maharishi effect." What is the Maharishi effect? It is an increase in quality of life and decrease in violent crime that occurs when approximately 1% of the population of any given geographical area meditates continually in such a way that all the practitioners in the group achieve pure consciousness (i.e., consciousness without any content) at the same time.

The explanation of the Maharishi effect suggested on Scholarpedia involves the following propositions:

(a) "Consciousness is the unified field."

(b) "Attainment of pure consciousness by an individual meditator injects 'a wave of coherence' into the unified field."

(c) "If many individuals put such waves of coherence into the field at the same time, the effect spreads and becomes so intense that other individual consciousnesses in the vicinity are affected (even though they have never experienced pure consciousness)."

(d) "These essentially broadcast effects act to lower individual stress, increase life satisfaction, and decrease violent crime over an unspecified but limited geographic area." (Scholarpedia 2020)

Using the quantum theory concept of wholeness, the work of Prof. Hagelin appears to indicate that we are all connected. The entire universe is joined through our collective consciousness. To sum up his work, let me cite some of his own words, reported here by Jeanne Ball:

While Superstring Theory and M-theory are undergoing important refinements, the consensus among leading theorists is that the unified field exists. "What we've discovered at the foundation of the universe is a universal field where all the forces and particles of nature are united as one," says Dr. Hagelin. "They are ripples on a single ocean of existence." (Ball 2017)

Stuart Hameroff and Roger Penrose's Quantum Consciousness Theory

The study of consciousness through QM takes us into uncharted territory. The concept of the wave–particle duality of matter is used to understand both the brain and consciousness. In QM, we learn that the wave function carries all the system's information, similar to how a hologram behaves. This is how, by applying QM processes to the brain and consciousness, researchers have found out that consciousness acts as a quantum field or appears to display similar properties to a quantum field.

From a young age, Prof. Stuart Hameroff was fascinated by issues surrounding consciousness, particularly by the feeling that microtubules may play a more significant role in understanding consciousness. Inspired by Allison and Nunn's work (see Allison and Nunn 1968), he looked closely at the role of microtubules and anesthetics associated with consciousness. The importance of microtubules in his early work is evident in many of his publications. For instance, in "Consciousness, Microtubules and 'Orch-OR': A 'Space-time' Odyssey," he informs us that "consciousness occurs due to quantum vibrations in the brain microtubules" (2014, 126).

After several years of working alone, Hameroff partnered with Sir Roger Penrose, a British mathematical physicist who won the Nobel Prize for Physics in 2020. He has published a great deal on mind, quantum theory, quantum computing, and the brain. Hameroff and Penrose have taught us a lot about brain information processing. Their research suggests that using quantum theory to understand the brain's functioning and the place of consciousness has been worthwhile. One of their joint work outcomes indicates that the brain behaves as a quantum computer and that the brain processes information through quantum mechanical processes.

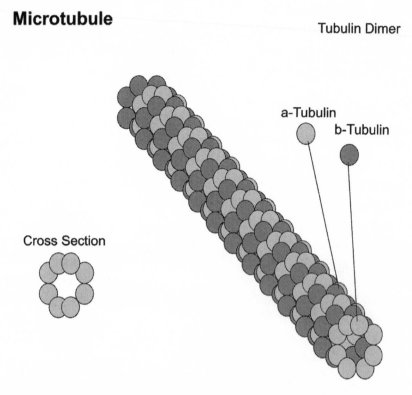

Figure 5.2 *The structure of a microtubule filament, which creates shape in a eukaryotic cell. Image by Lydiawc1020 via https://commons.wikimedia.org/wiki/File:Microtubule_ Structure. svg (Creative Commons Attribution-Share Alike 4.0 International)*

Their research suggests that consciousness arises from microtubules and activity inside neurons. This conclusion contradicts those who have put forward the argument that it occurs through the connections between neurons. In a nutshell, they have developed the orchestrated objective reduction theory, arguing that consciousness originates from quantum interactions in the microtubules inside each cell. Microtubules transport material inside cells. The theory posits that the tiny cellular structures, known as microtubules, produce consciousness.

Their work (Hameroff and Penrose 2014) suggests that there may be an eternal soul (consciousness) that does not perish after death. Hameroff argues that the human brain is a quantum computer and that the soul (or consciousness) is simply information stored at the quantum level. He adds that this information is stored after death and that quantum information (consciousness) is not destroyed but merges with our universe and exists for an unspecified period.

To sum up, their work uses quantum theory to explain the phenomenon of consciousness. Their joint research may have discovered consciousness carriers (taken here as information or elements that accumulate during one's life). At death, consciousness moves somewhere else. Where are these elements located? According to this research, these elements can be found inside protein-based microtubules, which function as carriers of quantum properties inside the brain.

Their theory has received many criticisms: many researchers, including leaders in the field of consciousness studies, have dismissed the Hameroff–Penrose model. The objections have primarily been based on the fact that quantum effects are extremely difficult to maintain in the laboratory. Therefore, the brain as a living entity is not an appropriate instrument for quantum effects over long periods of time. Characteristics of living things, such as their warmth, wetness, and noisiness, do not allow quantum effects to occur continually. However, despite these criticisms, Penrose continues to argue that quantum physics offers us a platform for understanding consciousness.

Pim van Lommel's Unified Field of Consciousness

The Dutch scientist Dr. Pim van Lommel has worked as a cardiologist for 26 years in various hospitals in Holland. During this time, he has witnessed many near-death experiences involving patients who have suffered cardiac arrests. His unique experiences led him to look for explanations and carefully analyze his data. Now, he has a new understanding of consciousness. His approach to consciousness is unique and worth exploring.

Dr. Pim van Lommel tells us that roughly 4% of people have experienced a near-death experience (NDE). Due to new and better techniques used during resuscitation, the survival rates have increased. He observed that patients who had cardiac arrests and recovered were able to present remarkably detailed descriptions of what took place during the time in which they were pronounced clinically dead. He set out to explain this phenomenon, to solve the riddle of how a clinically dead patient can remember! He posed the question, is this something to do with consciousness?

I have suspected for a long time now that one way to gain insights about the nature of consciousness is to look at NDEs. Several people have experienced an NDE, including experiences such as out-of-body experiences (OBE), cardiac arrest, clinical death, and brain injury. Others have experienced something similar during meditation. For instance, before dying, some people see everything they have done since birth, or the light at the end of a tunnel.

In several of his articles, Dr. Lommel has answered numerous fundamental questions in the quest to decode the workings of consciousness during various stages or types of NDEs. He points out a survey conducted between 1988 and 1992 involving 344 patients who had been through 509 successful resuscitations. The outcome of the study was published in *Lancet* in December 2001 (van Lommel et al. 2001). Several other studies have also been conducted by Bruce Greyson (2003) on NDE.

Dr. Lommel addresses essential issues and arguments relating to NDEs using QM and biology in a creative and exciting way. He puts forward clearly and concisely all the evidence for NDEs. His work suggests, and data confirms beyond doubt, the existence of nonlocal consciousness. He reports on what the research has uncovered. When you finish reading his book or some of his articles, you want to revise everything you ever learned in physics, biology, and chemistry. about human beings' true nature.

For instance, one realizes that consciousness is independent of the physical body, restricted neither by space nor by time. And through this view, one learns of the interdependency and interconnectedness that link and bind us with each other and with everything in the universe. He has gone deeper into researching consciousness than anybody else I have read. In my view, he has done far better than anyone before him in digging deeply, particularly into the problem of consciousness.

His analytical skills are prominently displayed in detailed accounts of many NDE cases, which feature profound insights rarely encountered in scientific books and articles. His use of concepts from quantum theory to look at the ideas of wholeness, interconnectedness, and nonlocality is a bonus for those without a background in physics. His work is well documented and supported by solid evidence. He says to us that consciousness is nonlocal. This means that the brain acts as a receiving instrument, along with other hardware in the body, such as cells.

Dr. Lommel's work on consciousness is by far the most exciting that I have come across. Anyone who wants to understand how consciousness works should read his book *Consciousness Beyond Life: The Science of the Near-Death Experience* (2010). There is no doubt that his writings will change the way people think about life, human beings, and animals, the way we interact with each other, and our purpose on this planet. Through his analysis of NDEs, Dr. Lommel has demonstrated that it is a real, distinctive, and valid form of experience that has nothing to do with psychosis, drug use, oxygen deprivation, or other causes to which doubters want to attribute this phenomenon.

He also points out that in many of the patients who have gone through NDEs, their personalities have undergone a profound and permanent change. Reading his work, one realizes that the overwhelming views and lines of reasoning held by the majority of those who work in the field of the brain or consciousness, mainly, those who claim that consciousness arises from the brain, are too narrow, incoherent, incomplete, poorly understood, and perhaps unreliable.

One realizes that we have this omnipresent, ever-present consciousness that exists, independent of space–time. Our consciousness has access to waves that carry information throughout. Ripple is the leading carrier of information, which manifests as images, pictures, voices, inspiration, and feelings. This consciousness, for instance, the sleeping consciousness, lets you have access to all knowledge, such as access to new dimensions, experiences, or ideas, during sleep. Our consciousness, being nonlocal, is timeless as well.

Dr. Lommel has used two physics concepts well-known in QM, namely non-locality and the idea of interconnectedness. The website Tmhome tells us, quoting Dr. Pim van Lommel: "The mind seems to contain everything at once in a timeless and placeless interconnectedness. The information is not encoded in a medium. Still, it is stored non-locally as wave functions in non-local space, which also means that all information is always and everywhere immediately available."

For those who support the view that consciousness arises from the brain, neurons, or some other physical stuff, I urge you to read this genius of consciousness. It will change your thinking about the brain and how you view consciousness. One learns the human mind's ability to go through experiences that are not understood by the overwhelming materialistic paradigm. Dr. Lommel's book provides excellent arguments on the existence and importance of consciousness. It is difficult and challenging to understand why and how mainstream science rejects the NDE implications concerning the overwhelming scientific work, the evidence presented by Dr. Lommel.

In addition, common and recurring characteristics by those who have experienced NDEs, particularly in the case of the OBE, are as follows:

- They no longer had a physical body.

- The experienced a feeling of being outside their dead body.

- They retained their consciousness; consciousness became a separate entity to the physical body.

- A review of 93 corroborated cases of verifiable OBE perceptions during an NDE showed that 90% were found to be completely accurate, while 8% contained a minor error, and 2% were shown to be incorrect.

- Conscious perception by the self is possible outside the body.

Furthermore, during a life review, the following were observed:

- During a life survey, time and space do not seem to exist; patients could instantaneously survey their whole life, hinting at nonlocality. And during this review, one felt an instantaneous connection with all those he or she knew and connected with during their lifetime

- Communication with dead relatives or friends was done instantaneously through "thought transfer."

- There was an interconnection between one's consciousness and those of other dead relatives.

Also, during resuscitation, consciousness return to the body:

- Patients have said that consciousness returned to their body, and most patients said it returned through the top of the head.

- Patients regained consciousness but felt that they are trapped in their physical, damaged, or crippled body.

- They came back because they had not finished their duty on this planet.

- They realized that there was self-consciousness as well as a field of consciousness.

- Consciousness is independent of our own body; it is a separate entity.

Similar work was done by Stephan A. Schwartz (2016), an accomplished author of several books. In his articles and publications, he points out that several million people worldwide have revealed experiencing an NDE. However, the number may be higher, as most people do not report it. Reading his works, one realizes that he has demonstrated that a growing body of theoretical and experimental work challenges the overwhelmingly dominant view that consciousness arises as a neurophysiological process; one realizes that a new paradigm, one that incorporates nonlocal consciousness, is developing.

What Dr. Lommel has discovered had been known for several years by some physicists. For instance, David Bohm discussed these issues of interconnectedness in several of his articles on QM. Looking at the universe closely, he pointed out that the universe behaves like a hologram, a metaphor that explains the concept of interconnectedness or oneness. This approach leads us to the idea of the unified field and in turn to the realization that this field is a field of consciousness, an intelligent field.

The following conclusion emerges from the work of Dr. Pim van Lommel:

- Individual consciousness has access to everything in the universe and is not restricted by time or space. It has access to all information.
- The wave functions store all aspects of consciousness in the form of information.
- The brain and the body are simply the receivers of consciousness.
- The brain does not produce consciousness.
- The brain function can be taken as a transceiver.
- The brain facilitates consciousness to be used by us, the body.
- Near death experiences have been experienced all around the world.
- Current theories and scientific views are unable to explain the causes and contents of NDEs.
- All past, present, and future events exist and are accessible (nonlocality).
- Consciousness retains memories and self-identify; it has a perception.

- Consciousness is there in a nonlocal space as a wave field of information.

- Consciousness can see, hear, and travel instantaneously.

- The brain is only an instrument of consciousness.

- The brain only serves as a relay station for parts of these wave fields of consciousness to be received.

- An analogy is made between a TV set, which receives electromagnetic waves and then transforms them into image, voice, and speech and the brain.

- As rightly pointed out by Dr. Lommel, "These waves hold the essence of all information but are only perceivable by our senses through suitable instruments, like the camera and TV set" (2011, 25).

- The brain is similar to a transmitter/receiver.

- Daily interaction exists between the invisible nonlocal space and our physical bodies.

- Consciousness can be experienced in another dimension, without the concept of space–time.

- Consciousness is a separate entity to the body.

- Our consciousness does not need the body to live, but our body needs our consciousness to interact and to live.

- Consciousness is nonlocal.

Summary

The purpose of this chapter has been to clarify the nature of consciousness and of its role in running the universe. It appears that consciousness is at the center of everything we do. There is a need to consider a new paradigm in physics that sees consciousness as primary. Several writers and scholars before me have discussed similar ideas: among these, David Bohm and Karl Pribham stand out. The latter put forward the idea that our brain behaves like a hologram, while the former argued that we live in a holographic universe.

This chapter's discussion suggests that consciousness is nonlocal and is the source of everything, the origin of everything that exists, the creative force running the universe. It seems likely that everything in the universe is shaped by nonlocal consciousness. It has become clear to me that our consciousness exists and predates our birth and the physical body. In addition, consciousness survives independently of the body after death, in a nonlocal, timeless, and spaceless environment, independent of both time and space. We live in a field or sea of consciousness, called a unified field of consciousness, where our consciousness exists both in our bodies and in this unified field.

Understanding consciousness from the field viewpoint leads without a doubt to the realization that consciousness continues to live beyond the biological death of the body. Our consciousness continues to live long after our physical bodies are destroyed. Important conclusions emerge from this book, including the existence of what I call nonlocal consciousness, which means that consciousness is independent of space–time and is not located in the body. Another conclusion is that individual consciousness is interconnected and interdependent.

The experimental and theoretical descriptions suggest that entanglement and nonlocality appear to be universal properties or attributes of our world and the universe. It is impossible to deny the causal role of consciousness. Its primary role has been established in this chapter. The chapter has also answered several fundamental questions regarding our physical world. Furthermore, it has provided profound insights into the nature of physics.

This chapter has built upon preceding chapters regarding matter, atoms, energy, and information to discuss consciousness, the source of everything in the universe. We have seen how unique physical laws govern the universe and its physics. We have seen how consciousness runs according to its mathematics and has its own logic. This investigation has led ultimately to the discovery of a unique underlying field called consciousness. The underlying field is an intelligent information system, the basis of our universe. This field is the source of everything.

There is something else I want to point out. It is the difference between how the proponents of the EM field theories and QM theories understand consciousness. The proponents of EM field theories of consciousness posit that consciousness arises when a brain produces an EM field with specific characteristics. In contrast, those leaning toward the QM theories suggest that consciousness exists in a field, independently of any human body or brain. Consciousness is a separate entity that can live without a physical apparatus. In a nutshell, they do not consider consciousness as an EM phenomenon.

It is regrettable that the approach discussed in this chapter, although well-known by many scientists, is discarded. What are scientists afraid of? Why do they have reservations about embracing the new view of consciousness? The approach used in this chapter is based on many discoveries in quantum physics and the fields of information and computation. The implications of looking at consciousness as primary can transform physics and the way we view matter, ourselves, and the entire universe.

From time to time, we do things automatically. We do certain daily tasks without even thinking because they have become routine. Our brain continually processes all these activities. However, from time to time, we make mistakes.

And it is precisely at the time that the error is made that something takes control. What is that something taking control? It is merely our consciousness or mind that takes care and supervision. We need to look at that something taking control of the brain or the body. That something is intelligent; that entity is consciousness.

I am aware that this work presents a challenge to established paradigms in physics and that many may not be easily persuaded. Hopefully, this study has laid the groundwork for future work. This chapter promoted peace, dialogue, togetherness, and the need to be interconnected. It promoted relaxation and called upon all of us to accept and embrace this message. Spread the message all over the world for the benefit of all humanity.

PART SIX

Conclusion

Consciousness is Fundamental

I regard consciousness as fundamental. I regard matter as derivative from consciousness. We cannot get behind consciousness. Everything that we talk about, everything that we regard as existing, postulates consciousness.

—MAX PLANCK

One of the aims and objectives of this book was to propose the idea that the running of the universe, matter, energy, atoms, electrons, and so on is grounded in consciousness. This argument lays a foundation for addressing many unresolved issues and unanswered questions in theoretical physics. I would like to reiterate that the aim was not to discuss consciousness per se. Much scientific research on consciousness has been conducted by several eminent researchers cited throughout this book.

For many years now, several theoretical physicists have shown a growing dissatisfaction with our most trusted physical model of the universe, mainly the SM of particle physics. In the SM, the universe is assumed to be made up of matter: 4% atoms, 20% dark matter (which we cannot yet observe), and 76% so-called dark energy. The SM of physics has many gaps. For instance, there is hardly any mention of consciousness in the model. More recently, there has been a growing interest in a new approach to understanding physics. Many scientists hope that a new approach to matter in which information and consciousness play an essential role may advance the field.

To stress once again, the overwhelming view that consciousness is produced by our brains and the idea that studying and learning about the brain, as neuroscientists do, will help us to uncover the location of consciousness and how the brain produces it has been shown to be inconsistent, despite several years of research and millions or perhaps billions of dollars of investment. So far, we have not been able to decode the working of consciousness through this approach, and definitely, the last word has not yet been said. Many experiments have shown a mismatch between consciousness and brain activity. For instance, it is well documented that in some parts of the brain, neurons associated with conscious experience have been located, while in other parts of the brain, no such experience seems to occur. This state of affairs suggests that brain processes do not produce consciousness.

The proponents of materialism or those who argue that everything is made up of matter and that it is primary have failed to account for consciousness's nature. The materialists have been unable to explain how consciousness emerges from animals or human beings. Many have concluded that materialism is untenable and that a new worldview is needed. Despite the work done in neurology, biology, and psychology, nothing has led to a deeper understanding of consciousness. We know quite a lot about our brain's mechanics—for instance, how the network of neurons in our brain works to compute and process information. But we have not yet discovered the mechanism whereby consciousness arises from the brain. It appears that the world of feeling and other forms of thinking belong to consciousness.

I propose that biologically and neurologically, the relationship between the brain and consciousness can be made clear by considering the brain (metaphorically) as a receiver, like the radio (transmission/reception). In this model, the brain does not produce consciousness but is instead a receiver. The brain experiences consciousness and transmits it into us. I am looking at the puzzle in this way to answer several fundamental questions. For instance, if the brain is damaged, consciousness is altered simply because the brain acts as a TV. If a TV (receiver) breaks down, its ability to receive or transmit information is impaired.

Furthermore, I firmly believe that one of the keys to unlock consciousness may be to look at several works that have been done in the field of OBE and to consider people who have had NDEs. It has also been noted that during OBE, shock, or NDE, consciousness moves outside the physical body and can experience and see what is going on in the physical world. This is supported by several experimental works by Dr. Van Lommel, indicating that consciousness is ontologically fundamental.

These experiences, documented by Dr. Van Lommel, tell us that the body is destroyed when it is no longer healthy at the time of death but that the soul or consciousness leaves the body and continues to live independently after death. Thus, we can argue that consciousness is deathless and remains for eternity; it never dies. Consciousness moves from one body to another and can live forever. This realization illustrates that consciousness is a separate entity and is fundamental. Without it in the body, there is no life. This discovery offers us some essential tools for possible solutions to the many problems found in physics, chemistry, and biology. It provides several answers to the many issues encountered by the proponents of the SM.

Many quantum theory experiments have hinted at the limitations of traditional methods to examine the nature of particles or subatomic particles. In a nutshell, some theories have become inadequate to the task of decoding the universe and answering several fundamental questions. The unexpected borderline between information, computation, energy, and consciousness has been encountered in many QM experiments, such as the Einstein–Podolsky–Rosen paradox and the double-slit experiment. For instance, in the latter, consciousness is shown to influence the functioning of the investigation and the outcome.

Many of these cited experiments suggest that consciousness is more fundamental than matter, particles, and energy. Physicists have realized that a new perspective is needed, one that encompasses information, computation, and consciousness. Experiments in QM show how minds affect the physical and tell us that consciousness, as a self-organizing process, transcends space and time. And those new laws of physics are needed. Established physical rules may not be able to explain consciousness. The reason is that consciousness is primary and non-epiphenomenal and that everything else is secondary.

Several QM theories of consciousness have recently become helpful in explaining consciousness and consciousness experiences. I have a firm conviction that the quantum theory approach will lead to a unified theory of consciousness. It appears that consciousness is central to the running of the physical world. Once again, during the double-slit experiment, our conscious experience affects the outcome of the investigation. Various studies have shown that consciousness is central. By avoiding looking closely at the nonmaterial imbued in physical objects, such as atoms, electrons, and the universe, the progress of physics is being held back—and so is humanity. Unfortunately, many leading physicists are not willing to embrace this new approach to physics.

Over the years, physicists have realized that despite various mathematical models and methods applied to resolve some of these issues, the limits of the methods used are becoming more apparent. That is why a new perspective or approach is needed. The essential ingredient of this new perspective was already pointed out by the pioneers of quantum theory, who assigned a unique role to consciousness. Fortunately, some physicists have realized the importance of the mind (consciousness) and information. We need to explore all possible avenues and see which one may be most useful. For instance, we need to remove the blockade to all progress by incorporating new concepts or metaphors and adjusting our theories.

There is no doubt that the links between consciousness and its associated factors, such as intentions, feelings, and emotions, are apparent. The evidence suggests that this is consistent with classical physics. Through several quantum theory experiments, there is evidence showing that the mind or consciousness influences matter. One consciousness talking to another— perhaps this is a case of two consciousnesses located in different objects interacting with one another.

We have learned from quantum theory that everything we thought was solid or physical is not physical. The quantum world appears to us like an invisible world because of our inability to see the tiny particles or fields with our naked eyes. We do not see the electrons or vortices of energy, which simply means that we have partial knowledge or are not aware of the true profound nature of matter. The electromagnetic spectrum (see Figure 6.1) teaches us plenty of lessons. We can only see less than one-ten billionth, meaning only less than 1% of the light spectrum. So, everything passing before us, through our bodies and the universe, is entirely invisible to our naked eyes. What does this mean? Simply, we are only presented with a small amount of reality, a small amount of what is happening around us—or saying it another way, we can only see part of the truth with our naked eyes.

We learn from physics that everything in the world is made up of matter, that matter is made out of atoms, and that atoms are composed of tiny particles, but these small particles are really vortices of energies. Matter registers information, such as color, voice, and image—in a nutshell, binary digits (bits) or quantum bits (qubits) of information. These vortices of energies behave like intelligent entities, embedded with consciousness. Or let us say simply that energy is made up of information and consciousness. When particles interact, they exchange information and get to know each other. They would interact as if they had minds of their own. Even if they are separate, they are connected because they share the same consciousness or are part of this intelligent field.

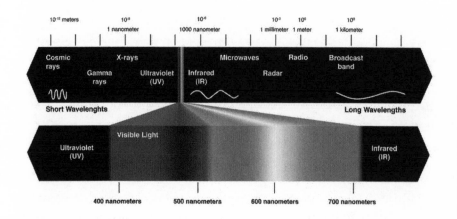

Figure 6.1   *Visible light of the electromagnetic spectrum. Image via https://www.pinterest. com/pin/ 502925483387631482/*

We are always broadcasting, emitting at a specific frequency. For instance, what two particles experience during entanglement may be similar to what a person experiences in meeting someone for the first time and falling in love. It is possible to argue that your consciousness (soul) feels attracted to this person and instantaneously recognizes them. There is a strong feeling of being part of the other and wanting to be together. Neither of the persons involved wants to be separated from the other. Both feel a special, unbreakable connection. It is this consciousness or soul within each of you that recognizes others. Each consciousness is imbued with information about every other—and possibly even information regarding past lives.

We experience the universal consciousness through shared pain, anger, joy, and happiness to sense what others are going through, such as our families, wives, children, and others. This is only possible if we are all connected by an underlying field of consciousness; without it, it would be impossible to feel or know what is going on far away. And it is for this same reason that entanglement between particles is possible. It follows that we are all holographically connected and that separation is an illusion.

Physics has all the ingredients, all the pieces it needs, to come up with solutions to the problems confronting it. It merely needs to connect everything to finally decipher the puzzle. Information and consciousness are vital pieces of this puzzle. And the change is already taking place. As rightly suggested by Shiva Meucci, "The complex, yet solid, 'crystal' of experimental evidence and mathematics we have developed in physics need not change dramatically and certainly shouldn't be destroyed or discarded! However, if we simply change the angle of incidence from which the light of interpretation shines upon it, the image that comes out the other side will change quite dramatically" (2020, 465).

It is clear now that consciousness is fundamental. Throughout the book, I have insisted on the role that consciousness plays in the universe's functioning. In the online article "The Puzzle of Conscious Experience" David Chalmers has proposed "that conscious experience be considered a fundamental feature, irreducible to anything more basic" (1995a). Putting consciousness at the center of physics—at the center of everything—will require a fundamental shift in attitude and the adoption of a new perspective. That is why I would like to point out that despite all the huge successes in physics over the years, despite the countless achievements in the fields of neuroscience and psychology, among others., the relationship between the universe and consciousness continues to pose problems. The question has remained so unanswerable because our approach has misled us into believing that consciousness is obscure.

Many solutions have been proposed—for instance, by David Chalmers. The solution to the hard problem of consciousness is elegant and straightforward. There is only one possible answer. It is to consider that consciousness is primary, the source of everything. It is the only thing required to explain physical features, not the other way around. This is the key to elucidating the relationship between consciousness and physical properties.

Throughout the book, I have shown that it is only through consciousness that the universe's physical properties can be understood. There is no doubt that recognizing the place that consciousness occupies in the universe will ultimately advance humanity and solve many of the riddles facing us. The concept of consciousness as primary is likely to answer several fundamental and complex questions in art, science, and religion and help us to understand consciousness itself as a separate entity—its properties, size, form, dimension, and other characteristics.

So far, there is not a single reliable theory that explains how matter can cause conscious experience. It is because everything, including matter, is a product of consciousness. That is why consciousness cannot be derived from matter. Yes, indeed, it is because consciousness is fundamental; that is why all the theories produced over the centuries by brilliant scientists have been unable to facilitate a physical theory of consciousness. Any idea of everything developed by physicists must incorporate consciousness. Otherwise, we are doomed to fail.

Throughout the book, I have pointed out how electrons are intelligent and imbued with consciousness. The SM of physics ignores consciousness. It was built up without any reference to consciousness, an essential and fundamental ingredient missing from physics. None of our leading physical theories have incorporated or even mentioned consciousness. That is why steps must be taken to integrate consciousness. Unfortunately, neither LQG nor ST has incorporated new discoveries from the science of consciousness.

However, on a positive note, the last few years have seen an increasing number of physicists embracing and incorporating the concept of consciousness. "Recently, particle physicists have made explorations into this challenging research arena that may initiate major paradigmatic shifts in cognitive neuroscience, resulting in the baton being passed to particle physics. Cross-pollination of ideas between particle physics and the cognitive sciences might help to develop a deeper understanding of consciousness, even though physicists' portrayal of the 'unphysical' is controversial" (Pandarakalam 2010, 1).

Subjective human attributes, such as pain, joy, anger, intelligence, happiness, and sadness, have nothing to do with the brain. There is no doubt that physics is entering a new phase, an original path incorporating information and consciousness. Physicists have realized that matter is energy. The "observer effect" teaches us that observing a particle affects the way the particle behaves. All these facts, taken together, insist on the primacy of consciousness over matter or energy. And ultimately, it leads us to the exciting realization that consciousness creates everything and manipulates matter, atoms, and electrons.

I hope this book helps to pave the way for a deeper understanding of consciousness and its importance in running the universe, including matter. I believe that by accepting consciousness as primary, we learn the true, intrinsic nature of matter. Hopefully, this understanding gradually but radically changes our approach to matter and the universe. Consciousness should be considered a creative force that shapes the universe, including matter, and constitutes the source of everything.

There is no doubt that the last few years have seen a vast interest in consciousness studies in almost all scientific disciplines, particularly in physics.

This trend is due to a growing dissatisfaction with the SM of physics. Physicists have identified a huge gap in the understanding of concepts ranging from the nature of matter to that of subatomic particles, along with QM issues, such as the double-slit experiment, entanglement, non-locality, and the mind–body problem.

A significant trigger for the surging interest in consciousness as an alternative to the overwhelming SM is that theoretical physics has reached a dead end. We are currently witnessing a radical shift in perspective, one that is leading us into uncharted territory. It is hoped that consciousness, as a fundamental framework, may deliver us from the challenges that theoretical physics has faced over the last 50 years. I believe that the idea of consciousness as primary has the potential to move the search for a ToE forward or at least point us in a more favorable direction.

I firmly believe that a new approach integrating the strengths of SM, ST, LQG, and consciousness is capable of addressing some of the most critical issues or difficulties associated with theoretical physics. I have shown that understanding physics from the consciousness viewpoint allows us to go beyond what we know about physics. This book's conclusion is that a general consciousness framework that can be used to tackle issues in theoretical physics is worthwhile.

By understanding consciousness as an intelligent field, we can now start to study this field scientifically. This means that we ought to learn about its properties, behaviors, and characteristics. We can then learn about the individual consciousness, about what it does during day and night. The individual consciousness is self-contained and simultaneously extends to the field. Look at NDE, OBE, and panpsychism. In *Digital Physics: Decoding the Universe*, I summarized some of the properties of consciousness. By doing this, we can quickly develop an empirical, scientific theory of consciousness. It is clear from this book that consciousness is fundamental and is the building block of the universe. It should be included in our models with other building blocks, such as energy, gravity, mass, space, and time.

Once consciousness is added, we can start to study matter, particles—everything in the universe—not only in terms of matter and energy but also in terms of information and consciousness. I discussed this aspect of consciousness in "A Special Relationship Between Matter, Energy, Information, and Consciousness" (Lokanga 2020). I stressed that in some cases, like in natural computers, information processing may be considered as similar to consciousness. I have stressed that QM has shown us that what we think of as a physical, solid object is, in reality, nonphysical. All I can say is that from an empirical view, consciousness is fundamental, the source of everything.

I can say now with confidence that the tendency to reduce consciousness to physical matter is untenable and has been shown to be unreliable. Those who have regarded matter as a fundamental entity of our world have been unable to tell us the origin of consciousness or its location nor have they been able to tell us where matter comes from. Physics does not tell us what matter, electrons, particles, and fields are; it only tells us what those entities do. It does not tell us anything about the intrinsic nature of fields, particles, or matter. The conclusion of this book is that these objects behave like intelligent entities, like consciousness itself. This realization leads us to argue that these objects (matter, atoms, electrons, trees, fields) may be considered as forms of consciousness.

In conclusion, let me stress once more that physics only tells us about electrons' behavior, not the intrinsic nature of matter. This book explains how consciousness is part of physics. It has laid the groundwork for physicists and for scientists in general to build a science of consciousness. Hopefully, the founding fathers of quantum theory will be delighted, wherever they are. They posited that reality was made up of two aspects, consciousness and matter, the mental and the physical, and that the two interact with one another.

Throughout the book, I have shown, using the approach of QFT, that subatomic particles are not physical objects, discrete entities, but rather fields. Particles are in a state of vibration, a movement pattern. An elementary particle is a vibration of an underlying quantum field. The particle is the field, continually moving. Using the perspective of QFT, we can learn the properties of this field—its size, length, breadth, speed, etc. This quantum field or electron field is what is fundamental, not the particle. Therefore, this underlying field is intelligent; it is consciousness itself. So, the unified quantum field is simply the universal consciousness.

PART SEVEN

Glossary of Scientific Terms

African binary system: The binary code, or binary notation, may have originated in Africa in the form of the Ifa divination that has been used in West
Africa, and throughout the rest of the continent, over the last 12,000 years—
which has led to the development of information technology.

Bantu: The Bantu people are the speakers of Bantu languages and are found
in sub-Saharan Africa. They believe that everything in the universe is imbued
with consciousness: they look at matter itself as a form of consciousness
(spirit or soul) because the soul uses it whenever it wants to.

Big Bang theory: The theory put forward that the universe originated as a
singularity, i.e., from the explosion of dense matter.

Bit: A bit stands for a binary digit and is the smallest unit of data in a computer. A bit has a single binary value, zero or one.

Block universe: The proponents of this theory argue that time is a four-dimensional space–time structure; time is like space. Each event or occurrence
has its own coordinates in space–time. It follows that nothing happens; there
is no change; past, present, and future are interwoven. All times are absolute
in every moment.

Casimir effect: The experiment conducted that showed that two mirrors
placed close to each other were attracted to each other. This force of attraction came to be known as the Casimir effect.

Classical field theory: A physical theory that predicts how one or more
physical fields interact with matter through field equations.

Complementarity: Niels Bohr put forward this principle in 1928. It says that
complete knowledge at atomic and subatomic scales requires two separate
descriptions, one of wave and the other of particle properties.

Computation: This is a mathematical calculation that includes both arith-
metical and non-arithmetical steps.

Computational theory of field: The author argues that the field performs computations and can be considered an intelligent system. And it is possible to suggest that we can learn a lot by studying how fields perform various operations in the universe and present us with multiple outcomes.

Conscious electromagnetic information (cemi) field theory: A theory put forward by Prof. Johnjoe McFadden using the concept of the electromagnetic field to explain consciousness.

Consciousness: The nonmaterial part of us is consciousness, soul, or spirit, and the body is simply the means through which consciousness expresses itself and experiences the world. It appears that consciousness is made up of an unknown substance, like a light.

Digital physics (DP): The physics of information, computation, self-organization, and consciousness.

Deoxyribonucleic acid (DNA): This is the complex chemical that carries genetic information. It is one of the constituents of all living beings. All information used to build a living being is encoded in its DNA.

Einstein–Podolsky–Rosen paradox: Albert Einstein produced many objections to the theory and argued that the present theory (quantum theory) was incomplete. This led him to formulate various thought experiments, such as the famous Einstein–Podolsky–Rosen paradox, a thought experiment designed to point out the inadequacies of QM. In 1935, a famous paper by Einstein, Podolsky, and Rosen examined the strange behavior of entanglement and concluded that QM was an incomplete theory. Einstein also made various attempts to refute the uncertainty principle.

Electric field: The physical field that surrounds each point in space when a charge is present.

Electromagnetic fields theories of consciousness: The electromagnetic theories of consciousness posit that consciousness can be understood as an electromagnetic phenomenon. Its proponents include Prof. Johnjoe McFadden, Tam Hunt, and Jonathan Schooler.

Entanglement or quantum entanglement: Separate particles are interconnected and are aware of each other's locations and movements, meaning that they can be treated as one unit or system.

Evolving block universe: The idea that space–time is best represented as an evolving block universe, in contrast to the block universe view.

Field: Its meaning in physics is associated with a physical quantity, represented by a number or tensor, which has a value for each point in space–time. Alternatively, it is a region in space–time where each point is under the influence of a force.

Frequency: Simply, the rate at which vibrations and oscillations occur.

Higgs boson field: One ingredient of the standard model of particle physics is a hypothetical quantum field responsible for giving particles their masses. This is known as the Higgs field. The particle associated with the Higgs field is known as the Higgs boson.

Hologram: A photographic recording of a light field used to display a three-dimensional image of an object.

Ifa divination: Ifa divination is a device for forecasting the future. The Ifa literary corpus consists of 256 possibilities called Odu.

Information: This word is not self-explanatory; it exists only as a potential and needs a medium to manifest. Hence, defining information is a difficult task. However, one idea we are all accustomed to is that information can be transmitted or communicated instantaneously. For instance, one can get direct information about a physical object's properties, such as a star in the sky or a tree located far away.

Instantaneous communication (nonlocality): Refers to action at a distance or the direct influence of one object on another distant object.

Interconnectedness: Merely the state of things being connected with each other and everything in the universe.

Large Hadron Collider (LHC): This is the world's largest and highest-energy particle collider and the largest machine in the world. It is located at CERN in Geneva.

Light spectrum: What, then, can we not see? The human eye has a limited ability to see beyond a specific spectrum. We can only see visible light. However, we learn from electromagnetism that light reaches us in various colors, such as infrared, ultraviolet, X-ray, and gamma-ray, which are invisible to our naked eyes.

Loop quantum gravity (LQG): This is one of the candidates for a quantum theory of gravity. Loop quantum gravity is based on classical general relativity. One of the essential conclusions of LQG is that gravity should be quantized.

Magnetic field: Refers to a vector field that describes the magnetic impact on moving electric charges, electric currents, and magnetized materials.

Microtubules: This component of the cytoskeleton is fibrous hollow rods that function primarily to help support and shape the cell. They are fundamentally important for maintaining cell structure and providing platforms for intracellular transport.

Near-death experience (NDE): This is a personal experience associated with death or impending death.

Out-of-body experience (OBE): Refers to a feeling of being separated from the physical body and of viewing it from far away, that is, from the outside.

Panpsychism: Panpsychism is the idea that everything in the universe is imbued with some form of consciousness and that consciousness is always there, associated with matter, and never emerges from somewhere else.

Quantum bounce: This is a hypothetical cosmological model for the origin of the known universe.

Quantum computer: This type of computer uses quantum mechanics to perform certain kinds of computation more efficiently than a classical or digital computer. Classical computers use a binary code, i.e., zeros and ones, while quantum computers can use more than two digits. In a quantum computer, information is treated and processed differently, using the entanglement of pairs of photons.

Quantum double-slit experiment or the double-slit experiment (DSE): This is one of the most famous experiments in physics. The experiment's outcome shows that light and matter can exhibit the characteristics of both waves and particles while also revealing the fundamentally probabilistic nature of quantum mechanical phenomena.

Quantum entanglement: Separate particles are interconnected and are aware of each other's locations and movements, meaning that they can be treated as one unit or system.

Quantum field theory (QFT): This is a theoretical framework that combines classical field theory, special relativity, and quantum mechanics but excludes general relativity's description of gravity. In QFT, particles are treated as excited states of their underlying fields. The fundamental objects of QFT are quantum fields.

Quantum gravity: This is a theory that attempts to develop scientific models that unify quantum mechanics (i.e., tiny things) with general relativity (very large things).

Quantum hologram: According to its proponents, this is a theoretical information-containing entity emitted by all physical objects above the molecular level and contains the entire history of those objects.

Quantum information: This is information about the state of a quantum system. It is the fundamental entity of study in quantum information theory and can be manipulated using quantum information processing techniques.

Quantum mechanics or quantum theory: Both refer to the same thing. It is a theory that attempts to explain physical behaviors at the atomic and subatomic levels.

Quantum tunneling: Particles can cross from one place to another without any restriction. Immune to any barrier, this is called quantum tunneling.

Quantum bit (qubit): In quantum computing, this is the basic unit of quantum information and is the quantum version of the classical binary bit that is physically realized with a two-state device.

Radioactive decay: Also known by various other names, such as radioactivity or nuclear decay, this is a process in which an unstable atomic nucleus loses electrons or undergoes fission.

Reductionist approach: In this book's specific context, the reductionist approach refers to some physicists' tendency to reduce physical phenomena to their most essential parts, i.e., breaking them down into smaller components. This approach is contrasted with holism, which looks at things as a whole.

Relativity: Encompasses two theories of Albert Einstein: special relativity and general relativity.

Self-organization: This is the physics of information processing in complex systems. Chris Lucas defines self-organization as the evolution of a system into an organized form in the absence of external pressures.

Sleeping consciousness: Consciousness is active both during sleep and when we are awake. Thus, we are endowed with not one but two recurring forms of consciousness: one we experience during the waking state and the other during sleep.

Space and time: These concepts that allow us to describe events in terms of where and when they happened. All the laws of physics, without exception, involve space and time. In physics, we describe events using two parameters, location and time.

Space: The collection of all idealized events—where an event is something that happens at a point in space and at a moment in time.

Special relativity: A theory that deals with conditions in which gravitational forces are not present.

Standard model (SM): In the SM of particle physics, particles are considered as point-like objects.

String theory: This theory replaces subatomic particles with strings. The strings are either closed loops or open. They vibrate in different ways, and the different modes of vibration give rise to all the different particles in the universe.

Subatomic particles: These are particles that are smaller than an atom. There are two types of subatomic particle—elementary particles, which are not made of other particles, and composite particles.

Theory of everything (ToE) or the grand unified theory (GUT): A hypothetical final, single, all-encompassing, coherent theoretical framework of physics that fully explains and links together all physical aspects of the universe.

Time: Time is what a clock is used to measure. This is the generally accepted definition of time. A watch is a device that reads time. That is why, in our modern world, we have divided time into the second, minute, hour, day, week, month, and year.

Universal consciousness: This universal consciousness is ever-present throughout the universe and manifests itself through algorithms, programs, laws, and codes (in humans, animals, plants, and the universe as a whole). This intelligent consciousness manages the universe and maintains its harmony, and the laws that govern the universe are constant, balanced, and neutral.

Vibration: The word originates from the Latin *vibrationem*. In this book, it refers to a phenomenon whereby oscillations occur.

Waking consciousness: Consciousness we experience during the waking state.

Wholeness: The universe and life itself are conceived as a unity in which everything is connected and constantly interacting. We must start to see the whole picture, embrace this spirit of wholeness, and understand that we are all inextricably connected through our consciousness.

Yoruba: The Yoruba people are an ethnic group that inhabits western Africa, mainly Nigeria and Benin.

Zero-point energy (ZPE): The central concept of ZPE is that so-called empty space is overflowing with energy. Any empty space between atoms, even between planets, is filled with ZPE.

BIBLIOGRAPHY

Abbott, M., and Van Ness, C. 1972. *Theory and Problems of Thermodynamics, by Schaum's Outline Series in Engineering.* New York: McGraw-Hill Book Company.

Al-Khalili, J., and McFadden, J. 2015. *Life on the Edge: The Coming of Age of Quantum Biology.* London: Black Swan.

Allison, A. C., and Nunn, J. F. 1968. "Effects of General Anesthetics on Microtubules: A Possible Mechanism of Action of Anaesthesia," *Lancet* 2: 1326–29.

Alvele, B. 2020. "Exactly What Is Time?" Accessed July 22, 2021. http://www. exactlywhatistime.com/.

Anderson, H. K., and Grush, R. 2009. "A Brief History of Time Consciousness: Historical Precursors to James and Husserl." *Journal of the History of Philosophy* 47 (2): 277–307.

Andeweg, H. 2016. "Everything Is Energy, Everything Is One, Everything Is Possible." *Accessed June 17, 2020. https://www.turnerpublishing.com/ blog/detail/everything-is-energy-everything-is-one-everything-is-possible/ arXiv:1310.4691v1 [quant-ph].*

Atmanspacher, H. 2020. "Quantum Approaches to Consciousness" In *The Stanford Encyclopedia of Philosophy* (Summer 2020 Edition) edited by Edward N. Zalta Accessed August 28, 2020. https://plato.stanford.edu/ archives/sum2020/entries/qt-consciousness.

Baksa, P. 2011. "The Zero Point Field: How Thoughts Become Matter." Accessed June 17, 2020. https://www.huffpost.com/entry/ zero-point-field_b_913831.

Ball, J. 2017. "Collective Consciousness and Meditation: Are We All Interconnected by an Underlying Field?" Accessed September 14, 2020. https://www.huffpost.com/entry/collective-consciousness-meditation_b_822288.

Ball, P. 2017. "The Strange Link between the Human Mind and Quantum Physics." Accessed August 28, 2020. http://www.bbc.com/earth/story/20170215-the-strange-link-between-the-human-mind-and-quantum-physics.

Barbieri, M. 2004. "The Definitions of Information and Meaning Two Possible Boundaries between Physics and Biology." *Rivista di biologia* 97 (1): 91–109.

Barbieri, M. 2011. "The Paradigms of Biology." Accessed August 5, 2020. *https://dactylfoundation.org/wp-content/uploads/2011/05/TheParadigmsofBiology.pdf.*

Barbieri, M. 2016. "What is Information?" *Phil. Trans. R. Soc. 374: 1–10. http://dx.doi.org/10.1098/rsta.2015.0060.*

Barbour, J. 1999. *The End of Time: The Next Revolution in Our Understanding of the Universe.* London: Orion Publishing Group.

Barbour, J. 2008. "The Nature of Time." Accessed July 22, 2021. http://www.platonia.com/ nature_of_time_essay.pdf.

Barnett, M. S. 2009. "Introduction to Quantum Information." Accessed August 12, 2020. *https://www.gla.ac.uk/media/Media_344957_smxx.pdf.*

Bates, J. M. 2006. "Fundamental Forms of Information." *Journal of the American Society for Information Science and Technology* 57 (8): 1033–45.

Becker, K. 2014. "Is Information Fundamental? Could Information Be the Fundamental 'Stuff' of the Universe?" Accessed August 4, 2020. *https://www.pbs.org/wgbh/nova/article/is-information-fundamental/.*

Berezin, R. 2014. "A Unified Field Theory of Consciousness: A New Paradigm." Accessed August 27, 2020. *https://www.psychologytoday.com/gb/blog/the-theater-the-brain/201406/unified-field-theory-consciousness.*

Bidinger, S. 2020. "Everything Is Energy and . . . Energy Is Everything." Accessed June 17, 2020. *https://www.practical-personal-development-advice.com/everything-is-energy.html.*

Block, N. 2009. "Comparing the Major Theories of Consciousness." Accessed August 27, 2021. https://www.nyu.edu/gsas/dept/philo/faculty/block/papers/Theories_of_Consciousness.pdf.

Bohm, D. 1981. *Wholeness and Implicate Order.* London: Routledge and Kegan Paul.

Bohm, D., and Hiley, B. 1993. *The Undivided Universe: An Ontological Interpretation of Quantum Theory.* London: Routledge.

Borowski, S. 2012. "Quantum Mechanics and the Consciousness Connection." Accessed October 5, 2020. https://www.aaas.org/quantum-mechanics-and-consciousness-connection.

Brockman, J. 1999. "The End of Time: A Talk with Julian Barbour." Accessed January 05, 2022. https://www.edge.org/conversation/the-end-of-time.

Brown, L. 2018. "Reality Is an Illusion: Everything Is Energy and Reality Isn't Real." Accessed June 17, 2020. https://hackspirit.com/illusion-reality-scientific-proof-everything-energy-reality-isnt-real/.

Butterfield, J. 2001. "The End of Time?" Accessed February 19, 2021. https://arxiv.org/pdf/gr-qc/0103055.pdf.

Calhoun137. 2014. "Time: The Various Meanings and Interpretations of Time in Theoretical Physics." Accessed June 8, 2020. *https://medium.com/@calhoun137/what-is-time-dee7f911eafa.*

Casimir, H. B. G. et al. 1948. "The Influence of Retardation on the London-van der Waals Forces." *Phys. Rev.* 73: 360–72.

Chalmers, D. J. 1995a. "Facing up to the Problem of Consciousness." *Journal of Consciousness Studies* 2: 200–219.

Chalmers, D. J. 1995b. "The Puzzle of Conscious Experience." Accessed June 8, 2020. http://consc.net/papers/puzzle.html

Chalmers, D. J. 1996. *The Conscious Mind: In Search of a Fundamental Theory.* Oxford: Oxford University Press.

Chinedu, I. 2019. "Matter Is an Illusion: Physical Reality Is Empty Space Buzzing with Energy." Accessed June 8, 2020. *https://medium.com/@chineduimoh/ matter-is-an-illusion-physical-reality-is-empty-space-buzzing-with-energy- 4fea9e23e0b6.*

Chopra. D. 2019. "Physics Must Evolve Beyond the Physical. *Activitas Nervosa Superior* 61: 126–129.

Conrad, G. 2000. "Time, Space and All of Us: Trilogy, Book 2." Accessed June 27, 2020. https://www.universalmedicine.co.uk/about/shop/books/ time-space-and-all-us-0.

Cook, G. 2020. "Does Consciousness Pervade the Universe? Philosopher Philip Goff Answers Questions about 'Panpsychism.'" Accessed June 17, 2020. https://www.scientificamerican.com/ article/ does-consciousness-pervade-the-universe/.

Dittrich, T. 2014. "The Concept of Information in Physics': An Interdisciplinary Topical Lecture." *European Journal of Physics* 36 (1): 1–38.

Dodig-Crnkovic, G. 2010. "Biological Information as Natural Computation." In *Thinking Machines and the Philosophy of Computer Science: Concepts and Principles*, edited by J. Vallverdú, IGI Global (Hershey, PA), 36.

Dodig-Crnkovic, G. 2011. "Dynamics of Information as Natural Computation." *Information* 2 (3): 460–77.

Dodig-Crnkovic, G. 2012. "Information and Energy/Matter." *Information* 3: 751–55.

Emerging Technology from the arXiv. 2019. "A Quantum Experiment Suggests There's No Such Thing as Objective Reality." Accessed June 8, 2020. *https://www.technologyreview.com/author/ emerging-technology-from-the-arxiv/.*

Falk, D. 2016. "A Debate Over the Physics of Time." Accessed June 27, 2020. *https://www.quantamagazine. org/a-debate-over-the-physics-of-time-20160719/.*

Folger, T. 2007. "Newsflash: Time May Not Exist." Accessed September 11, 2021. *https://www.discovermagazine.com/the-sciences/ newsflash-time-may-not-exist.*

Fries, P. 2022. "Mechanisms and functions of rhythmic neuronal synchronization." Accessed January 27, 2022. https://www.esi-frankfurt.de/research/fries-lab/

Fung, Han Ping. 2016. "Re: Is Time an Illusion?" Accessed April 27, 2020. https://www.researchgate.net/post/Is_time_an_illusion/57ce35b9404854a1745eea46/citation/download.

Gefter, A. 2008. "Is Time an Illusion?" Accessed February 05, 2020. https://www.newscientist.com/article/mg19726391-500-is-time-an-illusion/.

Gershenson C. 2012. "The World as Evolving Information." In *Unifying Themes in Complex Systems VII*, edited by A. A. Minai, D. Braha, and Y. Bar-Yam, 100–15. Berlin: Springer.

Giovannetti, V., Lloyd, S., and Maccone, L. 2015. "Quantum Time." Accessed November 07, 2020. *arXiv:1504.04215v3* [quant-ph].

Glattfelder, J. B. 2019. *Information—Consciousness—Reality: How a New Understanding of the Universe Can Help Answer Age-Old Questions of Existence*. Zurich: Springer.

Goff, P. 2019. "Science as We Know It Can't Explain Consciousness—But a Revolution Is Coming." Accessed October 5, 2020. https://theconversation.com/science-as-we-know-it-cant-explain-consciousness-but-a-revolution-is-coming-126143.

Goff, P. 2020. "Consciousness: How Can I Experience Things That Aren't 'Real'?" Accessed October 5, 2020. https://theconversation.com/consciousness-how-can-i-experience-things-that-arent-real-139600.

Görnitz, T. 2014. "On the Different Aspects of Time in the Fundamental Theories of Physics." In *Direction of Time*, edited by S. Albeverio and P. Blanchard Springer, 21–29. Heidelberg: Springer.

Grandpierre, A. 1997. "The Physics of Collective Consciousness." *World Futures. The Journal of General Evolution* 48:23-56.

Greyson, B. 2003. "Incidence and Correlates of Near-Death Experiences in a Cardiac Care Unit." *Gen. Hosp. Psychiatry* 25: 269–76.

Gruber, P. R., et al. 2018. "The Illusory Flow and Passage of Time within Consciousness: A Multidisciplinary Analysis." *Timing & Time Perception* 6 (2): 125–153.

Hagelin, J. S. 1987. "Is Consciousness the Unified Field? A Field Theorist's Perspective." *Modern Sci. and Vedic Sci.* 1 (1): 29–87.

Hagelin, J. S. 2016. "Consciousness is the Unified Field." Accessed September 14, 2020. https://transcendentalmeditationblog.wordpress.com/2016/10/25/consciousness-is-the-unified-field-quantum-physicist-john-hagelin/.

Hameroff, S. 2014a. "Consciousness, Microtubules and 'Orch-OR': A 'Space–time' Odyssey." *Journal of Consciousness Studies*, 21(3–4):126–53.

Hameroff, S., and Penrose, R. 2014b. "Consciousness in the Universe: A Review of the 'Orch OR' Theory." *Physics of Life Reviews* 11 (1): 39–78.

Hameroff, S., and Penrose, R. 2014c. "Reply to Seven Commentaries on 'Consciousness in the Universe: Review of the "Orch OR" Theory.'" *Physics of Life Reviews* 11 (1): 94–100.

Harris, A. 2020. "Consciousness Isn't Self-Centered." Accessed August 28, 2020. http://nautil.us/issue/82/panpsychism/consciousness-isnt-self_centered.

Herzog, C. 2010. "What is Matter?" Accessed June 9, 2020. http://www.privatedimension.at/is_matter_an_illusion.html.

Higgins, C. 2018. "There Is No Such Thing as Past or Future: Physicist Carlo Rovelli on Changing How We Think about Time." Accessed June 27, 2020. *https://www.theguardian.com/books/2018/apr/14/carlo-rovelli-exploding-commonsense-notions-order-of-time-interview*.

Hobson, A. 2013. "There Are No Particles, There Are Only Fields." *American Journal of Physics* 81 (3): 211–23. https://arxiv.org/abs/1204.4616.

Hobson, A. 2017. *Tales of the Quantum: Understanding Physics' Most Fundamental Theory*. Oxford UK: Oxford University Press.

Hunt, T. 2018a. "Could Consciousness All Come Down to the Way Things Vibrate?" Accessed September 14, 2020. https://theconversation.com/could-consciousness-all-come-down-to-the-way-things-vibrate-103070.

Hunt, T. 2018b. "The Hippies Were Right: It's All about Vibrations, Man! A New Theory of Consciousness." Accessed September 14, 2020. https://blogs.scientificamerican.com/ observations/ the-hippies-were-right-its-all-about-vibrations-man/.

Hunt, T. 2019. "Is Consciousness Just a Complex Electromagnetic Field?" Accessed August 27, 2020. *https://medium.com/@aramis720/ is-consciousness-just-a-complex-electromagnetic-field-9d4bf05326f0.*

Hunt, T., and Schooler, J. W. 2019. "The Easy Part of the Hard Problem: A Resonance Theory of Consciousness." *Front. Hum. Neurosci.*13 (378): 1–16.

Hunt, T. 2020. "Electrons May Very Well Be Conscious" Accessed August 27, 2020. https://nautil.us/electrons-may-very-well-be-conscious-9008/#:~:text=For%20example %20the% 20late%20Freeman, by%20 electrons.%E2%80%9D%20Quantum%20chance%20is

Hurtak, J. H., and Hurtak, D. 2011. "Understanding the Consciousness Field." *The Open Information Science Journal* 3: 23–27.

Information Philosopher. 2020. "Information." Accessed August 4, 2020.

https://www.informationphilosopher.com/introduction/information/.

Iyer, R. 1984. "Spirit, Mind and Matter." Accessed June 11, 2020. https://www. theosophytrust.org/886-spirit-mind-and-matter.

Jaffe, A. 2018. "The Illusion of Time." *Nature* 556 (7701): 304–05.

Kastrup, B. 2020. "Will We Ever Understand Consciousness?" Accessed October 5, 2020.

https://iai.tv/articles/will-we-ever-understand-consciousness-auid-1288.

Kay, L. E. 2000. *Who Wrote the Book of Life? A History of the Genetic Code.* Stanford: Stanford University Press.

Kennedy, E. J. 2014a. "The Nature and Meaning of Information in Biology, Psychology, Culture, and Physics: 2. What is Information?" Accessed August 3, 2020. http://science.jeksite.org/info1/pages/page2.htm.

Kennedy, E. J. 2014b. "The Nature and Meaning of Information in Biology, Psychology, and Culture, and Physics: 3. The Nature and Meaning of Information in Biology, Psychology, and Culture." Accessed August 3, 2020. http://science.jeksite.org/info1/pages/page3.htm.

Kennedy, E. J. 2014c. "The Nature and Meaning of Information in Biology, Psychology, Culture, and Physics: 4. The Nature and Meaning of Information in Quantum Physics." Accessed August 3, 2020. http://science.jeksite.org/info1/pages/c4_physics.pdf.

Kennedy, E. J. 2014d. The Nature and Meaning of Information in Biology, Psychology, Culture, and Physics: 5. Integrating the Nature and Meaning of Information." Accessed August 3, 2020. http://science.jeksite.org/info1/pages/c5_integration.pdf.

Kleiner, J. 2020. "Mathematical Models of Consciousness." *Entropy* 22 (6): 1–52.

Kowall, J. P., and Deshpande, P. B. 2016. "It's the Other Way Around: Matter Is a Form of Consciousness and Death Is the End of the Illusion of Life in the World." *Journal of Consciousness Exploration & Research* 7 (11): 1154–208.

Kuhn, R. L. 2015. "The Illusion of Time: What's Real?" Accessed June 27, 2020. *https://www.space.com/29859-the-illusion-of-time.html*.

Lambrecht, A. 2002. "The Casimir Effect: A Force from Nothing." Accessed June 17, 2020. https://physicsworld.com/a/the-casimir-effect-a-force-from-nothing/.

Lamoreaux, S. 1997. "Demonstration of the Casimir Force in the 0.6 to 6 mm Range." *Physical Review Letters* 78 (1): 5–8.

Lenda, P. 2013. "Does Matter Exist or Is It All Just an Illusion?" Accessed June 9, 2020. *https://www.shift.is/2013/04/does-matter-exist-or-is-it-all-just-an-illusion/*.

Linde, A. 2018. "Universe, Life, Consciousness ." Accessed September 9, 2018. https://static1.squarespace.com/static/ 54d103efe4b0f90e6ca101cd/t/54f9cb08e4b0a50e0977f4d8/1425656584247/universe-life-consciousness.pdf.

Livni, E. 2018. "This Physicist's Ideas of Time Will Blow Your Mind." Accessed June 27, 2020. *https://qz.com/1279371/this-physicists-ideas-of-time-will-blow-your-mind/*.

Lloyd, S. 2002. "Computational Capacity of the Universe." Accessed September 9, 2018. https://arxiv.org/pdf/quant-ph/0110141.pdf

Lloyd, S. 2006. *Programming the Universe: A Quantum Computer Scientist Takes on the Cosmos*. London: Jonathan Cape.

Lokanga, E. 2017a. *Digital Physics: The Universe Computes*. Washington DC: Euclid University Press.

Lokanga, E. 2017b. *Digital Physics: The Universe Is a Programmed System*. USA: CreateSpace Independent Publishing.

Lokanga, E. 2018a. Digital Physics: The Meaning of the Holographic Universe and Its Implications Beyond Theoretical Physics. USA: CreateSpace Independent Publishing Platform.

Lokanga, E. 2018b. *Digital Physics: The Physics of Information, Computation, Self- Organization and Consciousness Q&A*. USA: CreateSpace Independent Publishing Platform.

Lokanga, E. 2020a. *Beyond Eurocentrism: The African Origins of Mathematics and Writing*. USA: KDP.

Lokanga, E. 2020b. "A Special Relationship between Matter, Energy, Information, and Consciousness." *International Journal of Recent Advances in Physics* 9 (1/2/3): 1–13.

Lokanga, E. 2020c. "Toward a Computational Theory of Everything." *International Journal of Recent Advances in Physics* 9 (1/2/3): 13–32.

Lucas, C. 1997. "Self-Organising Systems FAQ." Accessed September 8, 2012. http://www.mountainman. com.au/news97_h.html.

Macdonald, C. 2010. "Implications of a Fundamental Consciousness." *Activitas Nervosa Superior* 52 (2): 85–93.

Maguire, G. L. 2015. "Does Time Exist?" Accessed October 06, 2020. *https:// larrygmaguire.com/does-time-exist/#genesis-content*.

Martin, S. 2020. "Universe is Alive: Consciousness Pervades through Cosmos Down to Smallest Atoms—Claim." Accessed June 17, 2020. https://www.express.co.uk/news/ science/1229293/universe-alive-consciousness-definition-meaning-universe-meaning-space-philosophy.

Matloff, L. G. 2017. "Stellar Consciousness: Can Panpsychism Emerge as an Observational Science?" *EDGESCIENCE* 29 (9): 9–14.

Maynard-Smith, J. 2000. "The Concept of Information in Biology." *Philos. Sci.* 67 (2): 177–94.

McFadden, F. 2002. "Synchronous Firing and Its Influence on the Brain's Electromagnetic Field: Evidence for an Electromagnetic Theory of Consciousness." *Journal of Consciousness Studies* 9 (4): 23–50.

McFadden, F. 2002. "The Conscious Electromagnetic Field Theory: The Hard Problem Made Easy." *Journal of Consciousness Studies* 9 (8): 45–60.

McFadden, F. 2013. "The CEMI Field Theory Gestalt Information and the Meaning of Meaning." *Journal of Consciousness Studies* 20: 3–4.

McFadden, J. 2011. *Quantum Evolution: Life in the Multiverse.* London: Flamingo.

Meijer, K. F. D. 2013. "Information: What Do You Mean?" *Syntropy* 3: 1–49.

Meijer, K. F. D., and Geesink, J. H. H. 2017. "Consciousness in the Universe is Scale Invariant and Implies an Event Horizon of the Human Brain." *NeuroQuantology* 15 (3): 41–79.

Meucci, S. 2020. "Physics Has Evolved beyond the Physical: A Reply to Valid Criticisms of the Crisis in Physics." *Cosmos and History: The Journal of Natural and Social Philosophy* 16 (1): 452–65.

Mills, D. 2018. "How to Better Define Information in Physics." Accessed August 3, 2020. https://www.physicsforums.com/insights/how-to-better-define-information-in-physics/.

Minford E. "E=mc2: Everything Is Energy—Why Do We Continue to Ignore the Energetic Truth?" Accessed June 17, 2020. https://www.universalmedicine.co.uk/articles/emc2-everything-energy-why-do-we-continue-ignore-energetic-truth.

Mohideen, U. et al. 1998. "Precision Measurement of the Casimir Force from 0.1 to 0.9 μm." *Phys. Rev. Lett.* 81: 45–49.

Moreva, E., Brida, G., Gramegna, M., Giovannetti, V. Maccone, L., and Genovese, M. 2013. "Time from Quantum Entanglement: An Experimental Illustration." Accessed June 27, 2020. *arXiv:1310.4691v1* [quant-ph].

NEWS. 2020. "At Nautilus: Electrons Do Have a 'Rudimentary Mind.'" Accessed September 14, 2020. https://mindmatters.ai/2020/07/at-nautilus-electrons-do-have-a-rudimentary-mind/.

Noyes, K. 2011. "The Holographic Universe: Is Our 3D World Just an Illusion?" Accessed June 9, 2020. *https://www.technewsworld.com/story/ The-Holographic-Universe-Is-Our-3D-World-Just-an-Illusion-72804. html?wlc=1309984273.*

Otap, L. 2019. "Time Might Be Nothing but an Illusion." Accessed June 27, 2020. *https://medium.com/predict/is-time-an-illusion-2ee143dd653a.*

Pandarakalam, J. P. 2010. "New Horizons in Consciousness Studies." Accessed October 5, 2020. https://www.rcpsych.ac.uk/docs/default-source/ members/sigs/spirituality-spsig/pandarakalam-new-horizons-in- consciousness-studies-revised.pdf?sfvrsn=712f3d14_2.

Parker, E. B. 1974. "Information and Society." In *Library and information service needs of the nation: Proceedings of a Conference on the Needs of Occupational, Ethnic, and other Groups in the United States*, edited by C. A. Cuadra and M. J. Bates, 9–50. Washington, DC: U.S. Government Printing Office.

Pepperell, R. 2018. "Consciousness as a Physical Process Caused by the Organization of Energy in the Brain." *Frontiers in Psychology* 9: 1–11.

Pockett, S. 2012. "The Electromagnetic Field Theory of Consciousness: A Testable Hypothesis about the Characteristics of Conscious as Opposed to Non-conscious Fields." *Journal of Consciousness Studies* 19: 191–223. Accessed August 28, 2020. http://ingentaconnect.com/ content/imp/ jcs/2012/00000019/F0020011/art00008.

Pockett, S. 2013. "Field Theories of Consciousness." Scholarpedia 8 (12): 4951. Accessed August 27, 2020. http://www.scholarpedia.org/article/ Field_theories_of_consciousness.

Ponte, V. D., and Schäfer, L. 2013. "Carl Gustav Jung, Quantum Physics and the Spiritual Mind: A Mystical Vision of the Twenty-First Century." *Behavior Science* 3 (4): 601–18.

Popper, K. R., and Eccles, J. C. 1977. *The Self and Its Brain*. New York: Springer International.

Popper, K. R., Lindahl, B. I., and Arhem, P. 1993. "A Discussion of the Mind–Brain Problem." *Theoretical Medicine* 14: 167–80.

Powell, S. C. 2017. "Is the Universe Conscious?" Accessed August 27, 2020.

https://www.nbcnews.com/mach/science/universe-conscious-ncna772956.

Pribham, K. H. 1991. *Brain and Perception: Holonomy and Structure in Figural Processing*. Hillsdale: Lawrence Erlbaum Associates.

Pribham, K. H. 2004. "Consciousness Reassessed." *Mind and Matter* 2 (1): 7–35.

Pribham, K. H. 2006. "Holism vs. Wholism." *World Futures* 62: 42–46.

Puthoff, H. 1988. "Gravity as a Zero-Point-Fluctuation Force." *Physical Review* A 39 (5): 2333–42.

Puthoff, H. 1994. "Inertia as a Zero-Point Lorentz Force." *Physical Review* A 49 (2): 678–94.

Raker, S. 2015. "Quantum Mechanics and Near-Death Experiences." Accessed September 10, 2020. https://medium.com/@stevenraker/ quantum-mechanics-and-near-death-experiences-72e495998d56.

Ramakrishnan, P. 2017. "Quantum Theory, Modern Scientific Approach to Spirituality and #90DaysOfHeartfulness Meditation." Accessed June 11, 2020. *https://thriveglobal.com/stories/quantum-theory-modern-scientific-approach-to-spirituality-and-90daysofheartfulness-meditation/*.

Rawlette, S. H. 2019. "What If Consciousness Comes First?" Accessed October 2, 2020.

https://www.psychologytoday.com/us/blog/ mysteries-consciousness/201907/what-if-.

Robitzski, D. 2019. "Quantum Physics Experiment Suggests That Reality Isn't Objective." Accessed June 8, 2020. *https://futurism.com/ quantum-physics-experiment-reality-objective*.

Rose, D. 1986. *Learning About Energy*. New York: Plenum Press.

Rosenthal, D. 2015. "Concepts and Definitions of Consciousness." Accessed August 27, 2021.https://www.researchgate.net/ publication/267934696_Concepts_and_Definitions_of_Consciousness.

Rossi, L. E., and Rossi, L. K. 2018. "An Integrated Quantum Field Theory of the Evolution of Life, Consciousness, Cognition and the Neuroscience of Neuropsychotherapy." Accessed August 28, 2020. https://www.thescienceofpsychotherapy.com/an-integrated-quantum-field-theory-of-the-evolution-of-life-consciousness-cognition-and-the-neuroscience-of-neuropsychotherapy/.

Rovelli, C. 2018. *The Order of Time*. London: Allen Lane.

Russell, P. 2006. "The Primacy of Consciousness." Chapter contributed to *The Science and the Reenchantment of the Cosmos: The Rise of the Integral Vision of Reality*, by Ervin Laszlo. Rochester: Inner Traditions Bear and Company. Accessed December 08, 2021. http://www.peterussell.com/SP/PrimConsc.html.

Scholarpedia, 2020. "Field Theories of Consciousness / Field Theories of Global Consciousness." Accessed August 27, 2020. http://www.scholarpedia.org/article/ Field_theories_of_consciousness/Field_theories_of_global_consciousness.

Shani, I., and Keppler, J. 2018. "Beyond Combination: How Cosmic Consciousness Grounds Ordinary Experience." *Journal of the American Philosophical Association* 4 (3): 390–410.

Shanta, N. B. 2015. "Life and Consciousness: The Vedāntic View." *Communicative & Integrative Biology* 8 (5): 1–11.

Shelquist, R. 2006. "Volume XI—Philosophy, Psychology and Mysticism. Part I: Philosophy. Chapter XI: Spirit and Matter." Accessed June 11, 2020. https://wahiduddin.net/mv2/XI/XI_I_11.htm.

Siegel, E. 2018. "Ask Ethan: Are Quantum Fields Real?" Accessed June 12, 2020. *https://www.forbes.com/sites/startswithabang/2018/11/17/ ask-ethan-are-quantum-fields-real/#74db8498777a.*

Sipfle, K. 2018. "The Nature of Fundamental Consciousness." Accessed October 5, 2020. https://www.researchgate.net/publication/331306527_The_Nature_of_Fundamental_Consciousness.

Smolin, L. 2004. "Atoms of Space and Time." *Scientific American* 290 (1): 56–65.

Sohn, E. 2019. "Decoding the Neuroscience of Consciousness." Accessed August 28, 2020. https://www.nature.com/articles/d41586-019-02207-1.

Stonier, T. 1997. *Information and Meaning: An Evolutionary Perspective*. UK: Springer.

Street, J. 2015. "The Illusion of Time and Space." Accessed August 27, 2021. http://divine-cosmos.net/illusion-of-time.htm.

Sullivan, J. W. N. 1931. "The Sixth of a Series of Interviews by Mr. J. W. N. Sullivan with Leading Men of Science in This Country and Abroad." Accessed January 11, 2013. http://organizedreligion. me/2013/11/07/i-regard-matter-as-derivative-from-consciousness-max-planck-the-observer-january-25th-1931/tage point of digital physics.

Sunderland, G. (2015) "What did Werner Heisenberg and Erwin Schrödinger think about the role of consciousness in quantum mechanics?", Accessed 22nd June 2020. Available at: https://www.quora.com/What-did-Werner-Heisenberg-and-Erwin-Schr%C3%B6dinger-think-about-the-role-of-consciousness-in-quantum-mechanics

Taylor, J. 2020. "Illusion of Matter." Accessed June 8, 2020. https://werdsmith. com/ genesology/OLTGUS94p.

Taylor, S. 2017. "The World Is Not an Illusion." Accessed June 9, 2020. https://www.psychologytoday.com/gb/blog/out-the-darkness/201704/the-world-is-not-illusion.

Taylor, S. 2019a. "Is Consciousness a Fundamental Quality of the Universe?" Accessed October 2, 2020. http://www.sci-news.com/othersciences / psychology /consciousness-fundamental-quality-universe-07291.html.

Taylor, S. 2019b. "Spiritual Science: How a New Perspective on Consciousness Could Help Us Understand Ourselves." Accessed June 17, 2020. *https://theconversation.com/spiritual-science-how-a-new-perspective-on-consciousness-could-help-us-understand-ourselves-116451*.

Tegmark, M. (2007) "WHAT ARE YOU OPTIMISTIC ABOUT?", Accessed 22nd June 2020. Available at: https://www.edge.org/response-detail/10655

The Daily Galaxy. 2019. "The Ultimate Mystery? Consciousness May Exist in the Absence of Matter." Accessed June 17, 2020. *https://dailygalaxy. com/2019/09/the-ultimate-mystery-consciousness-may-exist-in-the-absence-of-matter-weekend-feature/*.

Thomas, E. 2019. "Three Female Philosophers You've Probably Never Heard of in the Field of Big Consciousness." Accessed August 28, 2020. *https://theconversation.com/three-female-philosophers-youve-probably-never-heard-of-in-the-field-of-big-consciousness-126974.*

't Hooft, G. 2018. "Time, the Arrow of Time, and Quantum Mechanics." *Frontiers in Physics* 6 (81): 1–10.

Tmhome. 2013. "Unified Field of Consciousness." Accessed August 27, 2020. https://tmhome.com/news-events/unified-field-of-consciousness-onemany/.

Tmhome. 2016. "Study on the Maharishi Effect: Can Group Meditation Lower Crime Rate and Violence?" Accessed September 21, 2020. https://tmhome.com/benefits/study-maharishi-effect-group-meditation-crime-rate/.

Tmhome. 2019. "A Scientific Perspective." Accessed September 21, 2020. https://tmhome.com/uncategorized/scientific-perspective-quantum-physicist-john-hagelin-phd/.

Tong, D. 2009. "What is Quantum Field Theory?" Accessed June 12, 2020. *https://www.damtp.cam.ac.uk/user/tong/whatisqft.html.*

Turner, M. S. 2007. "Quarks and the Cosmos." *Science* 315 (5808): 59–61.

Ullman, M. 1999. "Dreaming Consciousness: More than a Bit Player in the Search for Answers to the Mind/Body Problem." Accessed December 15, 2018. https://siivola.org/ monte/ papers grouped/copyrighted/ Dreams/ Dreaming_Consciousness.htm.

Umpleby, A. S. 2007. "Physical Relationships among Matter, Energy and Information." *Systems Research and Behavioral Science* 24 (3): 369–72.

Vallabhajosula, S. 2009. *Molecular Imaging: Radiopharmaceuticals for PET and SPECT*. Berlin: Springer.

van Lommel, P., van Wees, R., Meyers, V., *et al.* 2001. "Near Death Experiences in Survivors of Cardiac Arrest: A Prospective Study in the Netherlands." *Lancet* 358: 2039–45.

van Lommel, P. 2010. *Consciousness Beyond Life: The Science of the Near-Death Experience*. London: HarperOne.

van Lommel, P. 2011. "Near-Death Experiences: The Experience of the Self as Real and Not as an Illusion." *Ann. N.Y. Acad. Sci.* 1234: 19–28

Vikoulov, A. 2019. "The Unified Field and the Quantum Nature of Consciousness." Accessed August 31, 2020. https://www.ecstadelic.net/top-stories/the-unified-field-and-the-quantum-nature-of-consciousness.

Volk, S. 2018. "Can Quantum Physics Explain Consciousness? One Scientist Thinks It Might." Accessed September 14, 2020. https://www.discovermagazine.com/the-sciences/can-quantum-physics-explain-consciousness-one-scientist-thinks-it-might.

Webb, R. 2016. "Metaphysics Special: Is Time an Illusion?" Accessed June 27, 2020. *https://www.newscientist.com/article/mg23130890-900-metaphysics-special-is-time-an-illusion/?cmpid&utm_medium=EMP&utm_source=NSNS&utm_campaign=metaphysics_part6&utm_content=IsTimeAnIllusion.*

Weil, O. 2020. "How To Use Quantum Physics to Find Your Sacred Energy." Accessed June 17, 2020. *https://blog.sivanaspirit.com/sp-gn-quantum-physics-everything-energy/.*

Wheeler, J. A. 1989. "Information, Physics, and Quantum: The Search for Links." In *Proceedings III International Symposium on Foundations of Quantum Mechanics,* edited by *Ezawa, Hiroshi* (ed.) ; *Kobayashi, Shun Ichi* (ed.) ; *Murayama, Yoshimasa* (ed. et al.) , 354–68. Tokyo: The Physical Society of Japan.

Williams, A. 2005a. "Inheritance of Biological Information—Part I: The Nature of Inheritance and of Information." *TJ* 19 (2): 29–35.

Williams, A. 2005b. "Inheritance of Biological Information—Part II: Redefining the 'Information Challenge.'" *TJ* 19 (2): 36–41.

Williams, A. 2005c. "Inheritance of Biological Information—Part III: Control of Information Transfer and Change." *TJ* 19 (2): 21–28.

Wilson, C. M. *et al.* 2011. "Observation of the Dynamical Casimir Effect in a Superconducting Circuit." *Nature* 479: 376–79.

Wolchover, N. 2016. "Quantum Gravity's Time Problem." Accessed October 15, 2021. *https://www.quantamagazine.org/quantum-gravitys-time-problem-20161201/#comments.*

Wolfram, S. 2002. *A New Kind of Science*. Champaign, IL: Wolfram Media.

Wysong, R. 2020. "Matter Is an Illusion." Accessed June
9, 2020. *https://www.asifthinkingmatters.com/
solving-the-big-questions-second-edition/34-matter-is-an-illusion*.

INDEX

Made in the USA
Columbia, SC
01 March 2023

13187335R00126